GALAXY

GALAXY

MAPPING THE COSMOS

JAMES GEACH

REAKTION BOOKS

For my family

Published by Reaktion Books Ltd
33 Great Sutton Street
London EC1V 0DX, UK
www.reaktionbooks.co.uk

First published 2014
Copyright © James Geach 2014

Printed and bound in China by 1010 Printing International Ltd

A catalogue record for this book is available from the British Library

ISBN 978 1 78023 363 5

Contents

The bright band of the Milky Way glows over Chajnantor Plateau in the Chilean Atacama, home of the Atacama Large Millimetre Array (ALMA). We are embedded within a great disc of stars, our cosmic habitat, which can be seen in this long-exposure image that includes a view of the dense, central bulge of the galaxy. Dark patches along the bright band betray the presence of obscuring interstellar dust, blocking out the light from stars behind.

ONE

Cities Beyond

I magine standing on a tall hill on the outskirts of a great city. Around you is a scattering of isolated settlements, sometimes nestled together into quiet hamlets. Looking citywards, laid out before you is a vast, glittering labyrinth of streets, parks and high-rises: a dense and sprawling conurbation, focused on a distant hub of glistening skyscrapers. But the most striking thing about this metropolis is its silence: you don't see another soul or hear any sirens; there is no distant urban murmur. The city appears to be eerily dormant. Yet it's all almost within reach, waiting to be explored. But for now you are isolated, marooned in the suburbs, and all you can do is gaze and wonder at the complexity and richness just beyond.

Turning your back on the city before you, the vista gives way to an expanse of flat, open country, stretching away to the horizon. Save for the occasional outlying village and town clinging to the suburbs, the city seems to be alone in an empty land. But the distant expanse lures your gaze. You strain your eyes, and see that there are a number of faint glints on the horizon. There are more in every direction you look, and you come to realize that the city is not alone, that the world must be much bigger than you thought, and that there might be other cities just like yours.

So it is with our own galaxy and other galaxies in the universe. This is a book about those galaxies: what we know and what we don't yet know. We live in *a* galaxy, of which the Earth, Sun and solar system are just a minute component, but the universe is teeming with other galaxies of different sizes and shapes. Our best estimate is that there are something like 200 to 500 *billion* galaxies in the universe. As we will see, many galaxies are like our own, but others are very different. The goal of an extragalactic astronomer is to understand how those galaxies came to be.

Perhaps the most extraordinary thing about galaxies is not the galaxies themselves, but the tremendous distances between them. In fact, only rather recently did humans determine that galaxies were self-contained entities separated by vast gulfs of space. Since this discovery, our understanding of galaxies, their formation and evolution has accelerated at an astonishing rate, so much so that we can now perform – and, more importantly, interpret – the most extraordinary experiments and measurements. We can detect the ripple-like fluctuations encoded in the relic radiation from the Big Bang that represent the very seed points for galaxy formation; we can observe the explosive deaths of stars in far-flung galaxies and track their fading brightness to provide information on the overall evolution and fate of the universe, as well as the evolution of the galaxies themselves; and we are now gearing up for experiments that aim to measure the cosmic signature of the moment that the very *first* stars formed in the first galaxies. We will touch upon some of these themes.

It is often said that we are in a golden age of research into the origin, evolution and fate of galaxies. It's remarkable to think that we, as a species, have only just become properly aware that other discrete star systems exist beyond the local collection of stars we call the Milky Way. Just as the stars you see in the night sky are almost unimaginably distant from the Earth, so the external galaxies are unimaginably distant again from the Milky Way. This is a view of the universe that was only experimentally confirmed in the early years of the twentieth century.

At first, we charted the galaxies nearest to us: the ones that, by virtue of their proximity, loom relatively large and bright in the sky. Aided by technology, and pushed forward by a deep, driving desire to understand the universe, a century later astronomers have now surveyed millions of galaxies, mapped their distribution in space, analysed their contents and measured their motions. We can now detect galaxies billions of times fainter, and in frequencies of light orders of magnitude beyond the biological capabilities of the human eye, the tool our ancestors relied on when our species first became curious about the contents of the sky.

But what is a galaxy? What are galaxies made of? How big are they? How did they form? Why are there different types and how have they changed over time? These simple questions form the bedrock of the field of galaxy evolution. We will explore these questions over the course of this book, but one thing that needs to be made clear right away is that many of these questions are still being answered. There are many mysteries waiting to be solved. That is what makes this field the most exciting in astronomy, and perhaps in all of science. There is the feeling of the frontier about it. I will try to bring you not only the cutting edge of observations and theory, but also an insight into the nuts and bolts of astronomical research. How is it done, what tools do we use and what do astronomers actually do day to day? To begin this journey, let's start at home, and the city before us.

Via Lactae

Look up into the sky on a clear, dark night, preferably when there is a new (that is, no visible) moon, and you are far away from any urban glow. Be patient. It takes a few minutes for your pupils to dilate, becoming better accustomed to the darkness and more adept in soaking up the faint illumination coming from beyond the atmosphere. Now use your eyes to scan from horizon to horizon. You will notice that the density of stars increases, and the sky brightens slightly, in a band that stretches across the sky. You are looking into the dense, starry plane of our galaxy, dubbed *Via Lactae* – Milky Way – by the first classical astronomers. We have taken our first step into galactic astronomy. The stars 'above' us are not distributed haphazardly in space, but organized into an ordered structure. In the case of our galaxy, that structure is a disc within which we are embedded. That glowing band you see is the light from billions of stars that the eye cannot resolve individually, but combined en masse into a diffuse glow, which is brighter where the disc is denser. If you look towards the constellation Sagittarius, you are peering into the very heart of the galaxy, the densest concentration of material: a bulge of stars that sits at the hub of the great disc.

Crossing the Milky Way at an angle of about 60 degrees, you may see another faint band of light, emanating from the horizon where the sun has just set, or is about to rise. This time we are seeing light emitted from another plane: the ecliptic, or orbital, plane of our solar system. This is called zodiacal light: sunlight scattered off myriad rock and dust particles trapped in the disc of the solar system. The angle of the ecliptic relative to the starry band of the Milky Way reflects how the orientation of the orbital plane of the solar system is tilted relative to that of the galaxy. A plane within a plane.

Our solar system is located about two-thirds of the distance from the hub of the galaxy to the outer edge, well away from the densest concentration of stars at the centre. The disc is not completely flat, so when we look in *any* direction away from the Earth we see the relatively nearby stars above

A panoramic image of the Milky Way in visible light, clearly showing the disc and bright (but partially dust-obscured) bulge. Ours is a large spiral galaxy.

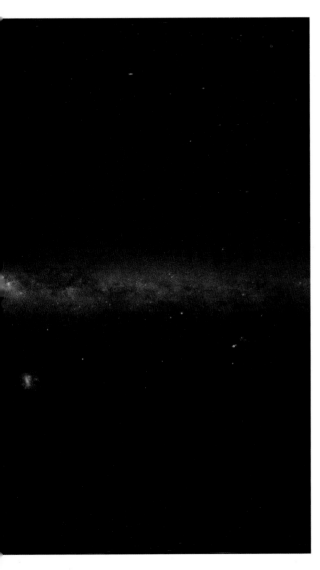

and below and all around us. Although they are all at different distances from the Earth, to our eyes the stars appear to be stationary points of light of various brightness on the inside of a huge sphere that surrounds the Earth. Indeed, this was the picture astronomers had for many years: these were 'fixed stars' on the 'celestial sphere'. Actually, on closer inspection, many stars actually appear to move across the sphere by small, but easily measurable, amounts, and we say that these stars have 'proper motion'. This is because they really are moving rapidly in space, and the signature of this is a year-on-year change in position on the sky that can be tracked with careful observation. For the casual observer, and on a human time-scale, stars generally do indeed appear fixed, but if you were to go to sleep and reawaken in a few million years' time, the constellations you would see would look different from the way they do today. The galaxy, and its contents, are in motion.

What the human eye cannot reveal is the three-dimensional distribution of the stars: they are scattered at various distances from us throughout space, not on the surface of some thin shell surrounding the Earth, as was once thought. It's important to note that most of the constellations are not physical associations of stars, but chance alignments of stars at different distances that happen to form recognizable patterns to us. An astronomer on a distant planet in some other region of the Milky Way would see a different set of constellations.

Some collections of stars *are* physically associated with each other. Binary systems are two stars in orbit around each other, and appear as a close pair on the sky, often so close that it is hard to separate them by eye, or with one star much brighter than the other, drowning out the companion (the bright star Sirius is an example of such a system). A large fraction of the stars in our galaxy are in binary systems. There are also larger groupings of stars, called clusters, which form because many stars can be born in the same place, bursting into life from the collapsing clouds of gas that are the crucibles of stellar matter in galaxies. A famous example is the Pleiades cluster, also known as the Seven Sisters, in the constellation of Taurus. The stars in the Pleiades formed fairly recently, are extremely luminous and in relatively close proximity to each other, making the Pleiades easily visible to the naked eye.

Scattered around the disc of the Milky Way, in an environment called the 'halo', we also find very dense balls of stars called globular clusters. Globular clusters contain hundreds of thousands of stars and are quite mysterious objects. The stars in each ball are held together by gravity, and the globular clusters themselves are gravitationally bound to the Milky Way, zipping about like flies around a dinner plate. The formation of globular clusters is still quite poorly understood, but they represent some of the oldest components of the galaxy, and therefore hold valuable clues about the formation of the Milky Way, and indeed other galaxies. A small telescope or binoculars will reveal some of the famous globular clusters, and they are among the most spectacular galactic sights.

The three-dimensional positions of the stars near to the Earth have been mapped using one of the oldest methods of distance measurement in astronomy: parallax. It's worth quickly explaining parallax, because the definition neatly provides us with the basic unit of measure for professional astronomers, the parsec. We will refer to the parsec later on when exploring the

The largest globular cluster in the Milky Way, Omega Centauri, a collection of 10 million stars in the 'halo' – the environment surrounding the disc – of the galaxy. There are about 200 known globular clusters in our galaxy, and they represent some of the oldest galactic components, although their origin is unclear. Omega Centauri might be the remnant of a dwarf galaxy that was accreted onto the Milky Way in the past. As such, it provides archaeological clues to the history of the formation of our galaxy.

vast scales of the other galaxies. It's not complicated: it's just a unit of measure encompassing a very large number of metres that would be too cumbersome to write down longhand, just as we don't measure car journeys in centimetres.

Close one eye and focus on the tip of your thumb held at arm's length. Now open the closed eye and shut the other. It looks as though the position of your thumb has changed relative to the background. This is parallax, and is simply the shift in the apparent position of an object when viewed along different sightlines. By knowing the difference in position of viewpoints, in this case the distance between your eyes, and the shift in *apparent* position of the target object, we can work out the actual distance via some simple trigonometry. Your brain is doing this constantly, and is in part what gives us depth perception. We don't notice depth in the star field the same way that we do in our local environment because all the stars are so far away that apparent changes in their positions are tiny. The trick *can* be done with stars, but for astronomical parallax measurements, we require much longer

separations of our sightlines and excellent, high-precision measurements of the stars' positions on the sky. As it turns out, nature has handed us a simple technique to do just this. Every six months we all change position by 300 million kilometres when the Earth is on the opposite side of the Sun during its annual orbital path. By observing the position of some distant star once, then repeating the measurement six months later, we can pull off the same thumb trick: the position of our eyes has changed. Of course, you don't have to wait the full six months, but this will give the longest possible baseline and the most accurate measurement of apparent displacement, and therefore the most accurate measurement of distance using this method.

When we measure the positions of stars, and indeed when we chart any object in the sky, we work in a coordinate system of angles. This relates back to the idea of a celestial sphere, a hypothetical huge screen enveloping the Earth onto which all the distant astronomical sources are 'projected'. Astronomers work in a similar system of latitude and longitude that is used on the curved surface of the Earth. This time, the grid lines are on the inside of the surface of that sphere (imagine standing at the centre of the Earth and seeing lines of latitude and longitude above you). These are called lines of Right Ascension and Declination. Just as locations on Earth can be described by pairs of latitude and longitude, we define the coordinates of objects on the sky in pairs of Right Ascension and Declination (or 'RA' and 'Dec.'), and the angular distance between any two coordinates in this system is the distance along part of a circle running along the inside of this sphere, called a Great Circle, with the longest possible separation being 180 degrees. As a guide, the width of the full moon takes up about half a degree on this sphere. Often astronomers will put a picture of the full moon next to large astronomical images for a handy angular scale comparison.

We can make finer divisions than degrees: like an hour can be divided into 60 minutes, one degree can be subdivided into 60 'minutes of arc', or arcminutes, and one arcminute can be divided into 60 arcseconds. One

Globular cluster 47 Tucanae (often referred to as 47 Tuc) is one of the most famous objects in the sky, visible with the naked eye in the southern hemisphere. It is imaged here in near-infrared wavelengths of light, and shows millions of stars concentrated into a dense ball. Remarkably, the full extent of this cluster is the same size as the full Moon on the sky, despite being around 350 billion times more distant. All these stars are held into their 'globular' configuration by gravity – they are orbiting a common centre of mass. In turn, this cluster is gravitationally bound to the Milky Way. All massive galaxies are surrounded by a posse of several hundred to several thousand globular clusters for very massive galaxies (such as ellipticals). The cluster 47 Tuc is a popular target for astronomers because it contains many interesting populations of stars. You will notice many bright stars that appear yellow/orange in this image. These are red giant stars: massive stars in a phase of stellar evolution where most of the hydrogen gas has been consumed, which are now burning helium, physically expanding to a super size in the process. Their red colour is related to their relatively cool surface temperatures (as far as stars go), of around 4,000 degrees. Betelgeuse in the constellation of Orion is an example of a red supergiant. Stars such as this provide windows onto a critical phase in stellar evolution.

arcsecond is about the same as the width of a strand of hair viewed from a distance of 10 metres. We can go further and divide more; in theory as much as we want, but in practice the smallest separations we can measure on the sky are set by instrumentation, which limits the accuracy, or resolution, with which we can pinpoint a given position. Those 'proper motions' of stars are often measured in units of thousanths of arcseconds, so you can see that these small changes in position are basically imperceptible to the unaided eye.

Now, consider a hypothetical star observed on this celestial screen. Imagine that we have measured its position once, and then wait six months and measure its position again, and work out the difference. The change in our physical viewing position is two times the Sun–Earth distance. If the apparent change in the star's position is two arc - seconds, then we define the distance to the star to be one parallax second, or one parsec (shortened to 'pc'). It's a pretty elegant unit of measurement, rooted in geometry. So parallax measurements are one way we can measure the true distance to stars, but since the apparent change in position gets smaller and smaller the further away a star is, there gets a point where accurate measurements cannot be made. In other words, parallax only works as a distance measure with-in a fairly small volume around us. You may be more used to hearing about distances in astrophysics mea-sured in light years, the distance light travels in one year

The Hubble Ultra Deep Field, a window onto the very distant universe. Nearly every point of light in this image is a galaxy, detected when Hubble gazed at a single, small area of sky (about 10 per cent of the diameter of the full Moon) for an extremely long exposure. Details in relatively nearby (but still very distant) galaxies can be seen, with clear spirals and ellipticals apparent, but the most distant galaxies are difficult to detect, appearing very small (sometimes just a few pixels across), faint and red in colour. Nevertheless, detecting these distant galaxies is essential to learn about the properties of galaxies when the universe was very young: the light from the most distant galaxies in this image was emitted when the universe was half a billion years old. We are looking into the past.

in a vacuum. Actually, except in some cases, extragalactic astronomers tend to use the parsec rather than the light year. It's more empirical because the definition is rooted in a geometric measurement. For comparison, though, one parsec is equivalent to just over three light years. The nearest star to the Sun, Proxima Centauri, is 1.3 parsecs away, and there are a few hundred stars within ten parsecs. The positions and parallaxes of over 2.5 million stars (and their proper motions) were mapped out with a European satellite called Hipparcos (an acronym for High Precision Parallax Collecting Satellite, but also a reference to Hipparchus, the ancient Greek astronomer), which operated between 1989 and 1993. The newly launched satellite Gaia is now performing a new survey charting the positions of a *billion* stars in the galaxy, providing the most accurate and complete view of the three-dimensional

layout of our cosmic habitat. Still, this only just scratches the surface of the total number of stars in our galaxy: Gaia will measure about 1 per cent of all the stars in the Milky Way. It's a bit like peeking out of the door and checking out where all the houses are in your neighbourhood, but still, Gaia is an incredible leap forward in this area. The galaxy contains many more stars than we can measure parallaxes for, and most of these are in the direction of that bright Milky Way band.

Look away from the Milky Way band and you are starting to look out above or below the galactic disc and towards truly deep space. Extragalactic space. Far beyond the nearby stars, and beyond the disc, there is a dark and silent chasm that contains more galaxies. Many more. Hundreds of *billions* more. Sadly, we never have a clear view of them, because in any direction we look, we're peering out through our own galaxy, which is full of stuff (not least our blazing Sun, which dominates our sky). Extragalactic astronomy is like being in a forest, standing underneath a giant oak and trying to see the trees in some distant woodland beyond. To observe all other galaxies, we have to look out in directions that point above and below the dense, starry disc of our own galaxy. The region of sky in the direction of the galactic plane is so bright and thick with intervening matter that it is effectively opaque to the light from the distant universe. To study other galaxies, we don't even bother looking in that direction. We call it the 'Zone of Avoidance'.

The Chandra Deep Field South is the name of an extragalactic survey region that has received significant observational investment from many different telescopes. It is known as a 'blank field'. The aim was to point at a part of the sky where there were no known points of interest (say, a large nearby galaxy), so that astronomers could conduct an unbiased, blind survey of a large number of galaxies seen at different 'redshifts'; that is, different cosmic epochs. This optical image is about the size of the full Moon, and has an exposure time of about two days. There is a scattering of stars as we peer out through our own galaxy, but the vista reveals myriad galaxies beyond. We are one of many.

The contents of a galaxy

So, our galaxy is a vast collection of stars organized into a disc-like structure. It's not immediately obvious to us that this is the case, because we as observers are deeply embedded within the disc itself. When we look out at the night sky we're just seeing the nearest stars to us and so don't get the full picture, just as you wouldn't appreciate the vast extent of the Amazon rainforest by standing in the middle of it. However, we can see the whole forest by studying other, distant galaxies that are far enough away for us to be able to see them in their entirety. To learn about other galaxies, we need to understand a bit more about what makes up a galaxy in addition to the stars. The Milky

Way, it shouldn't surprise us to learn, is a fairly average galaxy. A quick overview of its contents will set us up for our exploration of other galaxies in the universe.

We've already seen that the stars in the Milky Way are distributed in a disc. At the hub of this disc, there is a more spherical bulge of stars, called, well, 'the bulge'. If the Milky Way were a fried egg, the bulge would be the yolk. For reasons that we'll explore later, the stars in the bulge are different from the stars in the disc: they're older, on average. Within the disc itself, the stars are not smoothly distributed – there are higher-density regions that follow a spiral pattern similar to the spiral of a hurricane or snail shell, and this is where we find the youngest stars: new stars are forming in patches within in the spiral arms of the disc.

Just as the Earth orbits the Sun, the entire disc of the Milky Way rotates like a spinning plate, carrying the entire solar system on a galactic orbit. At the radius of our Sun, the rotation speed of the disc is about 200 kilometres per second, and it takes about a quarter of a billion years for us to orbit the hub of the galaxy once. So since its formation, the Earth has made nearly twenty full orbits of the Milky Way. As we will see more and more, our galaxy, and other galaxies, are dynamic entities, never at rest.

A wide view of the central region of the Milky Way showing the bright background stellar field smoky with interstellar dust, which is thickest in the plane of the disc. Here and there we can see patches of diffuse emission: blue light from young stars scattered and reflected by the gas and dust in their vicinity, and the red-pink glow of ionized hydrogen (HII) around the sites of formation of new stars.

What of the other contents of the galaxy? Among and between the stars is gas of various densities and temperatures, and this makes up the environment we call the interstellar medium. The gas is mainly hydrogen, the simplest, lightest and most abundant element in the universe, composed of a single proton and electron bound together. There are three main 'phases' of gas in galaxies: atomic gas, which is just gas composed of an ensemble of individual atoms; molecular gas, gas composed of ensembles of two or more atoms that have bonded together into molecules; and ionized gas, which is gas composed of atoms that have been irradiated or energized such that one or more of the electrons in the atoms has been stripped away. Although the majority of the gas in the interstellar medium is hydrogen, there are also other trace elements present: carbon and oxygen, for example (luckily for us). These trace elements were not present at the start of the universe but have been formed over time through the process of galaxy evolution; in particular the cycling of gas in generations of star formation.

In the disc of our galaxy the younger stars are distributed the way they are because of the underlying distribution of gas. The stars are born in giant clouds of molecular hydrogen, and one cloud can birth a generation of many stars. Since the gas clumps together under the influence of gravity, this gives rise to discrete patches of star formation throughout the disc of the galaxy, and specifically within the spiral arms, where the density of gas is highest. A star ignites when gravitational collapse pulls enough gas in close

concentration for a dense 'cold molecular core' to be formed. When the density of this environment is high enough, atoms within the core can fuse together, releasing vast amounts of energy when they do so. This is star formation. Many stars can be born at once as a collapsing cloud randomly fragments due to turbulence and other variations in the density of gas within it. A clutch of stars born close together becomes a cluster, gradually drifting away from each other over time.

Not all star clusters are globular. This is an 'open' cluster of stars in our galaxy called NGC 3603. In the disc of galaxies like our own, stars are born in giant clouds of molecular gas, which – during gravitational collapse – can spawn many stars in one go, resulting in clusters like this (although not all stars are born in clusters). Surrounding the cluster can be seen the glow of interstellar gas (the light is from the elements hydrogen, sulphur and iron), energized by the radiation emitted by this collection of young, massive stars. Understanding the exact physics of the details of star formation by studying regions of active stellar growth in our own galaxy provides vital information for the interpretation of star-forming galaxies in the distant (and early) universe.

After they are lit, the stars release the energy produced in their nuclear reactions in the form of ultraviolet and visible light. This radiation immediately impacts on the unburned gas still left in the birthing grounds, irradiating it with high-energy photons and creating bubbles where the gas becomes ionized, causing it to glow. This nebular glow can be a dead giveaway for identifying the sites of star formation in any galaxy. Ionization is the process where a photon of sufficient energy can eject an electron from an atom (or molecule). At some point that electron can return to the atom (or another atom that has also lost an electron), but to do so, it must release the energy *it* gained when it was ejected. It does this by emitting a photon. The quirk is that this 're-radiation' releases photons of a very specific energy (this is for quantum mechanical reasons: we can think of the electrons sitting in discrete energy levels around the atoms, like rungs on a ladder, and differences in these rungs correspond to very specific energies). The energy of a photon is directly proportional to its frequency, which we perceive as colour. Therefore, when new stars illuminate their natal clouds of hydrogen, the clouds glow with a very specific colour. We call the light H-alpha, and it has a red glow (with a wavelength of about 630 nanometres). These glowing regions are called HII regions, because HI is the shorthand for neutral (non-ionized) hydrogen, and so HII is the shorthand for hydrogen that has been ionized once. We'll come back to the different gas components of galaxies throughout the book.

In astronomy, all the elements other than hydrogen and helium (and, strictly, deuterium and lithium) are simply known collectively as 'metals'. The 'metallicity' of a region is a measure of how rich the region is in material other than primordial hydrogen and helium, and is usually quoted in units relative to the metallicity of the Sun. Where do the metals come from? It's become a dinner-party cliché to say that the Earth and everything on it, including us, is 'stardust' in the sense that we represent myriad permutations and reconfigurations of the ash of long-dead stars. It's true, of course, and nicely highlights a fundamental cosmic process. Stars are the alchemists of the universe; factories converting the base elements of hydrogen and helium into more complex forms in a process called nucleosynthesis. All of the elements we know about formed either during the nuclear fusion process that powers the star throughout its lifetime, or, for the heaviest elements (anything heavier than iron), in the extreme conditions that occur during the violent deaths of certain stars, called supernovae (explosive nucleosynthesis).

The gold in your ring formed in a stellar explosion (actually, it turns out, many different stellar explosions), and the carbon in your diamond was forged in the heart of a star. Don't have a diamond ring? The iron in your blood was formed in the same way.

As a result of this cosmic alchemy, aside from the metals that are produced, galaxies can contain large amounts of 'dust'. Dust is a general term used to describe the carbonaceous and silicate grain-like material (about the consistency of cigar smoke, but more diffuse) that is also created during stellar evolution. When a star dies, either by shedding its layers in a nova or explosively dispersing in a supernova, this dust is spread out into the interstellar medium. Dust tends to be accumulated in thick patches where the most active sites of star formation occurred, or are still occurring, and becomes obvious when we make large maps of the galaxy in visible bands of light.

The dust is generally opaque to the visible light photons, which get absorbed and scattered by the dust grains, just as it is hard to see clearly through a smoky room. The dust is better at absorbing and scattering photons of shorter, or bluer, wavelengths, so that preferentially the 'redder' photons make it through – this effect is called 'reddening'. Reddening means that in regions of high dust concentration optical observations alone will give us an incomplete view of what's going on, because a lot of the light is blocked out. The effect can easily be seen in the disc of the Milky Way. Any long-exposure image of the apparently starry plane will reveal darker patches and whirls within it. This is dust concentrated in the plane of the galactic disc, obscuring some of the light from stars behind it. To see through this dust, we have to turn to slightly longer wavelengths of light, as these can penetrate the dust more easily. Beyond visible light we have the 'near' infrared (because it is near the human-visible band of the electromagnetic spectrum), at wavelengths of one to a few microns. Near-infrared photons are not so susceptible to absorption by the dust, and so observing at near-infrared wavelengths provides an extra window through which we can see what's going on.

A 'stellar nursery' in our galaxy called IC 2944. The red light that dominates the background of this image is from ionized hydrogen. Every time an electron recombines with a hydrogen atom that has been ionized (that is, an atom that has had an electron 'ejected' when it absorbed a high-energy photon, emitted, say, from a nearby, young, massive, luminous star), light is emitted at a very specific wavelength. The exact wavelength depends on the energy of the transition: the rules of quantum mechanics tell us that the different possible energy transitions of electrons in atoms are discrete, like rungs on a ladder. One of the most common transitions in astrophysical environments where there is a lot of hydrogen gas and lots of ultraviolet radiation is called H-alpha; it has a wavelength of exactly 656 nanometres, in the red part of the visible light spectrum. This image shows the bright, young stars that are giving rise to the H-alpha emission, as well as dark blobs silhouetted against the background. These are called Thackeray's Globules, and are dense clouds of dust and gas. They are not easily penetrated by the optical light of the nebula that backlights them, and so appear dark. Slowly these globules are being evaporated as they bathe in the intense radiation field associated with those hot young stars, burning off the dust and dissociating the gas.

The Lagoon Nebula, a HII region in our galaxy. The red nebulosity shows ionized hydrogen, and the dark patches around betray the presence of a larger cloud of dense, dusty gas. This image covers an area about eight times the size of the full Moon on the sky. When looking at the discs of other nearby star-forming galaxies we can often see their spiral arms peppered with HII regions like this.

A view towards the centre of the Milky Way in the near-infrared part of the electro-magnetic spectrum, taken with the European Southern Observatory's VISTA survey telescope. Near-infrared light can penetrate the dust that obscures light in the visible part of the electromagnetic spectrum, providing a clearer view of the stars in the dense, crowded, blazing bulge of the galaxy. Nevertheless, in some places the dust is so thick that even the near-infrared photons cannot pass through, as can be seen in the dark filamentary structures that lace the image.

The dust can also emit its own characteristic glow, because, as it absorbs those ultraviolet and blue photons from stars, it heats up. This 'thermal' energy is re-emitted as infrared photons. Rather than the near-infrared, these photons tend to have much longer wavelengths, tens to hundreds of microns, in a region called the mid- to far-infrared. If you view the galaxy in these wavelengths, then the dust suddenly becomes bright, because it is the strongest emitter of those infrared photons. The stars themselves do not emit much radiation at these long wavelengths. Some of the most active galaxies in the universe – those that are forming the most stars – tend also to have the most dust, which blocks out the majority of the visible light coming from the stars, but the reprocessed infrared emission can dwarf all the other light coming from the galaxy, and so these dust-obscured galaxies can blaze brightly at infrared wavelengths.

A zoomed image of the centre of the Milky Way in near-infrared light, cutting through the dust that obscures the view of the central stellar population of our galaxy at optical wavelengths. The field is crowded with stars, and even here in the near-infrared we can see the telltale signature of obscuration, with the top left of the image darker, like a patch of spilled ink.

The products of stellar evolution have gone on to form other minor components of galaxies: asteroids, planets, plants, people. A quick aside: in the parlance of statistics, I think a vanishingly small number of astronomers would disagree with me that we expect life to exist in myriad forms on countless other worlds in other solar systems around distant suns in our own galaxy and in most other evolved galaxies in the universe. Frankly, it would take a lot of explaining if this was not the case. The range of complexity of these lifeforms probably forms a continuous distribution, from the cellular organisms growing in some sulphurous sludge on the moon of some distant planet, to the most technologically advanced civilizations that may have colonized many worlds, and perhaps inhabit the space in between. Perhaps it's reasonable to assume that we're at an average stage of development. That we haven't yet been contacted by other civilizations or detected a signal from some advanced society is most likely a reflection of the difficulty of intragalactic communication.

More fancifully, perhaps the politics of 'contact' among the more advanced societies of the wider galactic community is complex, or maybe even there is just a general apathy in this regard (which I find hard to believe, but I love

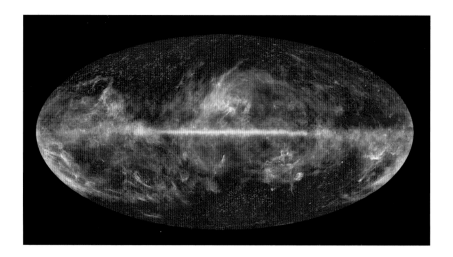

An image of the Milky Way taken by the Planck satellite. The goal of Planck was to study the emission of the Cosmic Microwave Background, but since it imaged the entire sky, Planck also provides us with detailed views of our own galaxy. This image shows in white-ish colours the wispy emission of dust in our galaxy, mainly concentrated in the galactic plane, but also revealing complex morphology above and below the plane (high 'galactic latitudes'). Most galaxies contain a large amount of dust, which is generated during star formation. This dust absorbs ultraviolet and optical light but emits radiation in the infrared part of the spectrum as it is warmed in the interstellar radiation field.

Iain M. Banks's novella *The State of the Art*, which describes a political *non*-contact scenario of Banks's 'Culture' with the Earth in 1977). The idea of life elsewhere in the galaxy, and indeed in other galaxies, is a fascinating and thought-provoking subject, but we're not going to dwell too long on this question. The Fermi paradox (also known as 'Where Is Everybody?' after Enrico Fermi's question) asks why, if the universe is so vast and abundant with habitable planets and presumably life, we haven't yet detected extraterrestrial civilizations. We'll answer this here with the sweeping statement that we will simply assume that life is out there, both in the Milky Way and other galaxies, and that it ranges from simple cellular forms to complex, technologically advanced societies with capabilities beyond our imagination. Whatever the case, it's worth remembering that complex biological systems are the eventual outcome and product of the process of galaxy formation and evolution.

Starting the journey

The Earth is bathed in the glow of the universe, the combined light emitted throughout the history of the cosmos. Our aim is to capture some of this light and understand where it came from and how it was emitted. This is the way we study galaxies. Cosmological experiments, in combination with the latest models for the architecture and evolution of the universe, point to the age of the universe being about 14 billion years. This is not a book about the Big Bang or cosmology (that is, the properties of the universe as a whole), but rather what has taken place in the universe since its formation that has shaped the most obvious outwardly visible characteristic of the cosmos: the galaxies. During this journey, we will learn how we think those galaxies formed and evolved, and will try to cover the main challenges and questions facing extragalactic astronomers today: when did the first galaxies form? How do galaxies grow and what physics controls a galaxy's fate? But more than this, we will explore a little bit how astronomy is done.

Throughout this book, we'll assume that the 'standard model' of cosmology, or 'concordance cosmology', holds. This is the model that describes the overall content, structure and evolution of the universe, and is called Lambda-CDM. Since the late 1990s it's generally been accepted by most astronomers as the best model we currently have to explain most of the observations of the overall properties of the universe. Lambda simply refers to the symbol for 'dark energy' that first appeared as the so-called 'cosmological constant' in Einstein's equations for general relativity. CDM stands for 'cold dark matter' (the exact nature of dark matter is not critical for our story; however, it will be a feature). Although we describe Lambda-CDM as our standard model of cosmology, both dark energy and dark matter are beyond our current standard model of *physics*, in that we do not know their true nature. This slightly disconcerting issue, combined with the fact that, in the concordance model, dark energy and dark matter combined make up the vast majority of the energy density, and therefore mass, of the universe, draws many critics to Lambda-CDM.

I don't want to get bogged down in cosmology, because this is a book about galaxies, and one thing is indisputable: Lambda-CDM is a model that successfully describes a wide range of empirical data. Of course, the nature of the cosmological model *is* certainly important in the history of the galaxies, but in some sense it is a distinct problem. I want to convey what we do know about galaxies through careful observation and analysis of data.

When describing light – that is, electromagnetic radiation – I refer to the frequency and wavelength of the light interchangeably (and deliberately inconsistently). Sometimes this is due to conventions in the particular sub-field of astronomy (radio versus optical, for example), but mainly it is to get across the point that describing the energy of light by its frequency is exactly the same as describing it by its wavelength – both mean the same thing, but refer to different aspects of the nature of electromagnetic radiation. If it helps, you can use the simple wave model of light, in analogy to a set of waves in the ocean: if you sit in a boat, anchored in an ocean swell, the rate at which your boat goes up and down from the passage of waves is the frequency. The distance between peaks or troughs is the wavelength. You can see how these two measurements are linked. Although the wave model is one way of describing the propagation of radiation, with light manifested as oscillations of a 'sea' of electromagnetic waves, I will usually describe light in the quantum model, where light is transmitted by photons. In either case, the frequency and wavelength of the light in question (be it radio waves, visible light or x-rays) correspond to the *energy* of the electromagnetic radiation.

There are two main things I want to get across in this story. First, there are many different *types* of galaxy in the universe. Second, these galaxies were not always around; they have changed – astronomers use the term 'evolved' – over the course of the universe. Naturally, there are areas where we lack under-standing, and for that we must turn to theory, models and speculation which we can compare to the observations. I make no apologies that some of the examples I will draw on are biased to my personal research experiences, and I admit from the outset that the scope of the subject is so broad that, by necessity, I have had to gloss over, oversimplify or omit some themes. But I want to give you some impression of what it's like to be an astronomer, how we perform observations and experiments from a practical standpoint, and how this has led to our present level of understanding.

With any luck, by the end I will have convinced you what a diverse, rich and exciting field of research this is, and shown you the remarkable things we have learned as a species over a surprisingly brief period of our history.

TWO

Stepping into the Extragalactic Universe

O ur galaxy is one of billions. Rather than all looking roughly the same, there is a rich variety in the types of galaxy that are out there. There are indeed many galaxies like our own spiral Milky Way, so called 'grand design' spirals (which is simply a shorthand way of describing the complexity of these objects) with different levels of spiral structure (some are wound tighter than others, for example). There are spiral galaxies with a linear 'bar' going through the hub; small, irregular and amorphous galaxies; merging and interacting galaxies with contorted shapes; and finally the massive galaxies with no discernable disc, but just a huge spherical or elliptical congregation of stars. These galaxies have different chemical compositions, contain different types of star and are forming new stars at a variety of rates. The Milky Way forms the equivalent mass of a few Suns in new stars every year; the most active galaxies can form new stars at rates hundreds of times faster than this, and some galaxies don't form any new stars at all. The galaxies are also not randomly distributed in space: they cluster together in a pattern determined by the underlying, omnipresent gravity. Different types of galaxy tend to be distributed in different ways. The goal of my research is to contribute to our understanding of why galaxies are the way they are and how they came to be. Why are there as many galaxies as there are? What processes have shaped them, and how has this changed over the history of the universe? Everything we know about galaxies today has come from patient decades of passive observations of the sky, as we have gradually collected the data from which we can glean important clues to answer these questions.

It should be becoming clear that there is something that the different components of galaxies have in common: they all emit light of some form or another, or, in some cases, block out light. If we can detect those different forms of light, or detect the 'absence' of light, then we can map out our

Face-on 'grand design' galaxy Messier 74, providing a spectacular view of the different components of a spiral galaxy, orientated face-on to our vantage point. The bright white-yellow-orange core glows with the light of billions of stars that are on average older than the blue (younger) stars seen throughout the disc. Patches of red are the brushfire-like HII regions where hydrogen has been ionized by new stars in giant clouds of molecular gas, the same as we see in our own galaxy. Laced through the disc and tracing the spiral arms is the dark signature of interstellar dust, the ashes of ancient stars.

The heart of the Whirlpool galaxy, another beautiful spiral galaxy orientated face-on to us, revealing the intricate structure of these 'island universes'. The two spiral arms emanating from the core are particularly bright with HII regions and thick lanes of dust. Note how the location of the dark dust and sites of active star formation are correlated.

A spectacular view of the spiral galaxy NGC 4921 in the Coma cluster of galaxies. The arms in this galaxy are not as pronounced as in other spirals (such as the Whirlpool), probably because of the fact that NGC 4921 is not forming stars as rapidly in its disc. This may well be related to the dense cluster environment: some of the gas required for star formation might have been removed from the disc during the galaxy's passage through the hot, dense intracluster medium. The background reveals myriad distant galaxies of many different types; they appear much smaller and fainter than NGC 4921.

The Black Eye galaxy, M64, an example of a spiral galaxy that is particularly awash with a thick lane of dust at its centre, obscuring much of the starlight.

NGC 1300, a barred spiral galaxy. The prominent spiral arms emanating from the ends of the linear bar structure are lit by young, hot, blue stars and star-forming regions, whereas the central regions are noticeably redder, indicating that the stellar population is, on average, more mature. Just visible in the very central region of the galaxy is another spiral structure representing a nuclear disc of gas and stars where star formation is also occurring. The bar structure is in part responsible for transporting gas and stars from the disc to the centre of the galaxy, as it redistributes angular momentum. Bars are important in shaping the formation of the bulges of some galaxies.

Galaxy NGC 5584, a 'flocculant' spiral galaxy ablaze with new, young, blue stars forming in patches throughout the disc. Flocculant spirals have a less well-defined spiral structure than so-called 'grand design' spirals: as in this case, they can be characterized by a clumpy, almost diffuse, structure, although the semblance of the spiral morphology can still be seen.

The central region of the spiral galaxy NGC 2841, beautifully contrasting the bright, smooth yellow-orange core – the hub of the disc – with winding spiral arms tattooed with dust, and punctuated here and there by blue clusters of new stars. As observers, we are embedded within the disc of a galaxy similar to this, looking out through the stars, gas and dust to the extragalactic universe beyond.

A wider view of the Whirlpool galaxy, M51. The large spiral is interacting with a smaller, more irregular galaxy appearing at the tip of one of the spiral arms. The gravitational interaction of the passage of this companion (which may well have passed through the disc of M51) is likely to have had an influence on the spiral arms, which are extremely well defined in this example, bright with new stars and HII regions.

A near-infrared view of the barred spiral galaxy NGC 1365. The formation of bars
– structures joining the spiral arms to the central hub – is due to gravitational
perturbations in the overall rotation of the spiral disc, causing some stars' orbits
to become very elongated, forming the bar. Up to two-thirds of spiral galaxies
today contain bar structures, including our own galaxy.

galaxy and others and break them down into their constituent parts. That's how we learn about the universe: we cannot observe or measure the material in a tactile way; we are totally reliant on detecting – or not detecting – photons that are emitted, absorbed or reflected.

If you think about it, the same is true in everyday experience: unless you go and actually put your hand on something, you either have to see it or hear it to know that it's there. In the case of eyesight, you're seeing photons that are either being emitted from an object directly (say, from a light bulb) or reflected from some object back into your eyes, like the scattering by dust motes of a shaft of sunlight, or your reflection in a mirror. Or, in the case of hearing, an object is somehow vibrating the air; perhaps the wings of a buzzing insect, making pressure waves that travel to your ear, vibrate your eardrum and are converted to sound in your brain. In both cases, you're not actually in physical contact with the object, but can learn something about its properties from the transmitted radiation. I can tell that grass is green, and therefore learn something about the biology of grass, without touching it, for example.

We started out just using our eyes, which were eventually aided by telescopes. Our eyes are sensitive to radiation with a fairly narrow range of energies, or frequencies, which we perceive as colour. The range of these frequencies roughly corresponds to the peak range of electromagnetic radiation that is emitted by the Sun and reaches the surface of the Earth. This is no biological coincidence; our eyes evolved so that we could 'see' this radiation, which is clearly a powerful evolutionary advantage. But the visible light is just a fragment of a continuous electromagnetic spectrum of radiation, emitted by different processes all over the universe. We have already learned how light with slightly longer wavelengths than the light we can actually see with our eyes can be used to penetrate the interstellar dust, for example. Today we have developed instruments we can attach to telescopes that can detect radiation right across the spectrum, ranging from gamma and x-rays (very high frequency, high energy) to radio waves (low frequency, low energy). Galaxies emit photons of every sort and, as we'll see, only by measuring them all can we be sure we have conducted a full 'census' of the astrophysics that governs the nature of galaxies.

Astronomy as an empirical science is unusual because we cannot conduct controlled experiments in the same way that 'laboratory' scientists do. Instead, from our limited vantage point – this one coordinate in the universe that we

Messier 104, or the Sombrero galaxy, is one of the most famous galaxies in the sky because of its distinctive morphology: a nearly elliptical, smooth bulge of stars combined with a very clearly defined dusty disc. We see the Sombrero nearly edge-on, but the disc is very slightly inclined to our view.

NGC 4710 in the Virgo cluster of galaxies is an example of an edge-on view of a spiral galaxy, so we can see the typical thickness of the disc and the round central bulge. Like most spirals, NGC 4710 has a clear dust lane concentrated in the mid-plane, obscuring and 'reddening' the central regions of the galaxy, just as in the Milky Way. Still, the prominent yellowish bulge of stars at the hub of the disc is clearly visible, as is the stellar 'envelope' of stars, which forms a diffuse white glow around the bulge and disc.

Another famous galaxy, the massive elliptical galaxy Centaurus A, imaged in optical light. Cen A is characterized by the warped dust lane snaking across its face.

The lenticular type Spindle galaxy, seen edge-on in this Hubble Space Telescope view. The extended white glow is the combined light of billions of stars, a large fraction of which are in the central bulge that dominates this galaxy. The prominent dust lane traces out the mid-plane of the disc, almost completely obscuring the light behind. Although lenticular galaxies are generally 'passive' – no longer forming stars – the dust was produced earlier in the galaxy's life when it was more active, and so hints at the past evolution of the Spindle. The formation of lenticular galaxies is not completely understood, but one possibility is that they have evolved from certain massive spiral galaxies.

A close-up view of the centre of Centaurus A, thick with interstellar dust, but fringed by new blue stars and the glow of HII regions. Centaurus A is an active galaxy that is undergoing a starburst and central black hole growth.

Elliptical galaxy NGC 1132, a beautiful illustration of the smooth, unblemished, spheroidal morphology of these massive, ancient galaxies. Ellipticals are generally 'passive' – no longer forming new stars. They underwent most of their stellar mass assembly earlier in the history of the universe, when the overall rate of galaxy formation was much higher than it is today. Ellipticals tend to be found in the densest environments – groups and clusters of galaxies – and are likely to have undergone significant merger activity in the past. Dotted around NGC 1132 can be seen thousands of globular clusters, appearing here as small pinpricks of light around the stellar envelope. As always, the background is filled with other, more distant galaxies: the cities beyond.

call Earth – we simply have to gaze out and soak up as much light as possible. Encoded within that light is the history of the universe. Astronomy is a bit like archaeology. Archaeologists can't just go and ask Julius Caesar what he had for breakfast; that has to be determined from other evidence. In fact, the parallels are a little deeper than that. Just as archaeologists look back into the past, so do astronomers when they look into deep space. The reason is that the distances between other galaxies and us are so huge that light from them takes an appreciable amount of time to get here. Hence the origin of the unit 'light year': the distance light travels in one year. So, just as when you hear a thunderclap you are hearing the boom emitted at the time you saw the lightning, which could be some seconds ago, when we look at galaxies we're seeing them as they were when they emitted their light. This can be billions of years ago, because the distances are so immense. The same effect is true of the light from our Sun. It's far enough away that it takes light eight minutes to traverse the inner solar system to Earth. So when you look up at the Sun (shielding your eyes, of course), you see it as it was nearly ten minutes ago. When you look at the Moon, you are looking back about one second into the past. Of course, light takes a finite time to traverse *any* distance, and so this time delay even applies to everyday life; it's just that the speed of light is so large in comparison to the human distance scale that we don't notice it.

This time delay provides astronomers with a handy tool: we can actually see what is (or rather, was) happening in the early universe simply by looking at distant galaxies. It is literally like having a window on the past. The purpose of extragalactic astronomy is not only to audit the contents of the universe, but to see how these have changed over time, and construct a physical model to understand it all. But how is astronomy actually done?

Observational astronomers throughout history, almost without exception, have been interested in one thing: collecting photons. We are light hunters. Those photons are the only direct connection we have to the distant universe, they having travelled for perhaps billions of years from a distant star or gas cloud to Earth, mostly unimpeded, but occasionally being absorbed and re-emitted or otherwise transformed or deflected along the way. Encoded within this steady rain of light is all the information we have to go on in order to understand the history of the cosmos. Unfortunately, given the tremendous distances involved, the amount of energy arriving at the Earth from any one galaxy is almost unimaginably – one might even say vanishingly – small.

The problem is compounded as we try to look further: galaxies appear smaller and fainter, and are hard to detect. To make matters worse, the tiny signal – the small portion of light hitting our detectors that is an iota of the whole – is drowned among a blazing sea of electromagnetic contamination, both natural and artificial: sunlight, street lights, radio transmissions and the infra-red glow of water in the atmosphere. To this end, astronomers have had to become ever more wily in their strategies for capturing this precious commodity, and the techniques they use to distil it into something meaningful.

It all boils down to two essential pieces of equipment: a telescope that can capture and focus light, and a detector to record it. The cutting edge of our science has always been honed through the development of the biggest telescopes and the most sensitive cameras and detectors. Unfortunately, astronomical instrumentation is both complex and expensive, and inevitably becomes more so over time. Most professional astronomers like me don't perform our research with the telescopes connected to our institutes; these are too small, and generally the sites are too poor (in terms of the weather) to perform the sensitive observations we require. Instead they are used for teaching purposes. In order to do the latest research, astronomers club together in multinational consortia, pooling funds and expertise to build giant telescopes and the cameras attached to them. There are only a limited number of geographic locations that can host such facilities; put the best (optical) telescope in a place where it's cloudy most of the time and you will have wasted your time and money.

The best telescope sites for astronomy are usually high, dry and suitably distant from populations so as not to be contaminated by the light of civilization. Of course, there is always the option to put telescopes into space, the most famous being the Hubble Space Telescope, but that's another financial and technical ball game altogether. The best sites for astronomy on Earth include the 4,000-metre-high peak of Mauna Kea on the Big Island of Hawai'i, which hosts many of the finest telescopes in the world; the high, dry Chilean Atacama which hosts the European Southern Observatory and the new Atacama Large Millimeter Array; and even the South Pole, which is another remarkably dry place.

There are lots of astronomers in the world, myriad astronomical targets, comparatively few telescopes and only a finite amount of time every year in which to conduct observations. So how do we actually get to do anything? The solution is simple: for each telescope, astronomers compete for the time

As the Earth rotates the 'fixed stars' describe a trail across the sky. This long-exposure view shows the Very Large Telescopes atop Cerro Paranal, dwarfed by the celestial sphere. Simply by staring out into the void we can explore the universe, its contents, and try to understand how it came to be.

they want to spend observing by writing a short proposal outlining what they want to observe and why. These proposals are submitted to a 'time allocation committee' (or TAC) – a group of our academic peers who review and grade each proposal based on the quality of the science goal and the feasibility of the experiment, and then dole out the precious time for each facility accordingly. If you want to take a snapshot of the Moon using the 8-metre Very Large Telescope in Chile, chances are you won't get time to do that project. But if you're a leader in the current hot topic, proposing some interesting new experiment with the promise of a high-impact result, then you might be lucky enough to be awarded some time. If you want to do something really wacky (interesting, but with a high risk of failure), then you might be asked to scale back your project and conduct an initial pilot study to see if your experiment is feasible: observe one galaxy instead of ten and see what you find, and then come back to us next year. Of course, the most sought-after telescopes are the biggest ones in the most exquisite sites with the best instruments, and these are always highly over-subscribed. The faction of astronomers that do not (usually) use telescopes – the theorists and simulators – can have a similar problem. While not using telescopes directly, most of their work is done on supercomputers. Theorists like big, powerful computers that can do the largest, most complex calculations; such machines aren't your garden-variety desktop PC, so often theorists must vie for time on shared-resource, high-performance computing facilities in a similar fashion to observational proposals. I hasten to add that these machines do not have to be located in such exotic and remote locations; just a climate-controlled room with an uninterruptable power supply.

Writing a telescope proposal is slightly different from writing a scientific paper. What we're doing is trying to sell an idea: making a project sound exciting and novel, while also being cautious and conservative; not being too greedy with the request, but asking for enough time to actually do something useful. It's a razor-edge balance. In general, telescopes (or, like the European Southern Observatory, groups of several telescopes) split the year into two semesters, and have a couple of 'calls for proposals' each year. By tradition, astronomers tend to leave it right to the deadline of submission before writing their proposals, resulting in frantic scrambles at the last minute, trying to put illustrative figures together, double-checking exposure time calculations and technical details, and honing the text into something that might appeal to the TAC. If we're lucky enough to be awarded time, typically

The mirror and camera of the 4-metre VISTA survey telescope, situated on Cerro Paranal, next to the European Southern Observatory's Very Large Telescopes in the Atacama Desert, Chile. One of VISTA's main goals is to perform very large surveys of the sky in near-infrared wavelengths of light, where it can detect large numbers of very distant, and therefore early, galaxies. Key to its success is the very large-format CCD camera VIRCAM: light collected by the 4-metre mirror is redirected and focused onto the camera to create large images of the night sky.

given in chunks of hours, or nights, then we can do what all astronomers crave to do: collect photons and learn about the universe accessible beyond our normal senses, getting a view that no human has ever seen. For me, that sense of discovery is where the excitement of this field lies.

After travelling across the universe, often for durations of time much longer than the existence of the Earth, the fate of a handful of the photons that were emitted from distant galaxies is to be captured by a mirror and focused onto a detector. This is the *raison d'être* of the telescope. Telescopes have grown and continue to grow bigger because of our insatiable desire to be better at collecting those precious photons. The bigger the mirror that collects the photons, the more light we can capture, and this means being able to detect ever fainter, ever more distant galaxies. When we talk about the light emitted by a distant galaxy, we refer to it in two ways. There is the galaxy's luminosity – the total amount of energy actually released by the galaxy every second (which is a very large number). And then there is the light we actually detect: that galaxy's observed 'flux' (which is a very small number). Why is the observed flux small? The flux is simply the energy we actually intercept here on Earth, and represents a minute fraction of the total luminosity of the galaxy.

Picture a distant galaxy as a 60 watt light bulb (or its nearest energy efficient equivalent). That light is being released in every direction, or 'isotropically'. Now imagine constructing a sphere that surrounds the light bulb. We will make the sphere opaque apart from a small cut-out square; let's make it 1 cm by 1 cm for neatness. Now, because the emission is isotropic, the flux shining (or, more correctly, flowing) through this square can be determined from the total wattage of the light bulb and the radius of the sphere. That 60 watts is spread out over the surface of the sphere, and the larger we make the sphere, the more spread out it gets. The flux through the little square can be calculated by working out the ratio of the cut-out area to the total surface area of the sphere. Since the luminosity of the light bulb (the 60 watts) is constant, if we make the sphere larger, then the flux through the cut-out gets smaller. In practice, the flux drops according to the inverse square law: make the radius twice as large and the flux drops by a factor of four; make it four times as large and it drops by a factor of sixteen. You can see, on astronomical scales those huge luminosities drop to small 'observed fluxes' rather quickly. But if you measure the flux, and know the distance to the source, or have some estimate of the distance, then you can use the inverse square law to calculate the intrinsic luminosity, which is generally what we want to do, as it tells us something about the nature of the galaxy: is it forming lots of new stars, for example?

In reality, let's go ahead and replace the light bulb with a distant galaxy and aim our telescope at it. Now, instead of a sphere built around the light bulb, imagine that the Earth sits on the surface of a huge invisible sphere that surrounds that faraway galaxy, and there is a flux of energy – photons – flowing through this sphere from that distant source. Our job is to capture some of this light using the mirror of a telescope. The problem is that the area of our telescope is tiny compared to the total area of the imaginary sphere, so we can only intercept a frighteningly small fraction of those photons. It is for this reason we need bigger and bigger telescopes.

In fact, as has been mentioned, there are only a handful of telescopes in the world capable of performing the observations we want, which are always fainter and farther – pushing the boundary of the observed universe. Not only do these telescopes have to be large, as we have seen, but we more often than not have to put them in extreme locations: generally the tops of mountains or high plateaus (or better, in space). This is because, after a journey of up to 10 billion years, those photons must get past a final barrier before hitting our detectors: Earth's atmosphere. This is full of nuisance molecules that absorb photons, and the problem is exacerbated for certain frequencies of light. The atmosphere is like a filter that blocks out some of the light coming from space. Take ultraviolet photons, for example; these are a very useful astrophysical probe because they are emitted by young, massive stars. The intensity of the ultraviolet (or UV) light of a galaxy can therefore be used to trace its active star formation (although this is a complicated issue). However, the Earth's atmosphere is exceptionally good at absorbing UV photons. Thank goodness, because this is what protects us from deadly levels of irradiating UV rays from the Sun. But this also makes UV astronomy from the ground exceptionally difficult, and only lets us get down to a wavelength of about 300 nanometres before all UV light is blocked. Put a UV-sensitive detector in space, above that atmosphere, however, and the problem is solved. A recent UV satellite, GALEX – the Galaxy Evolution Explorer – was launched in 2003 and came to the end of its mission in 2013. GALEX's goal was to measure the UV light emitted by young massive stars in local and distant galaxies in order to census the history of star formation in the universe. It performed observations that are simply impossible from the ground.

Not only does the atmosphere absorb some of the light we want to collect, but it also affects the direction in which those photons are going. This results in skewed, blurred and distorted images, like trying to focus on a coin at the bottom of a swimming pool. The principle in play here is refraction: the bending of the path of a light ray when it passes from one medium to another and changes speed. The atmosphere is not smooth and uniform, but made up of many different moving and turbulent layers and 'cells'. Instead of a coin at the bottom of a swimming pool, if you try to focus the point-like emission of a star, you will not see a stable, bright peak, but a blurred-out version that appears to dance around. The amount of blurring that occurs due to the atmosphere is called 'seeing', and until quite recently it placed a fundamental limit on the sharpness of images of astronomical sources that could be made from the ground.

There are two solutions to this problem. Option one is the simplest: put your telescope in space, so you don't even have to look through the atmosphere. The con is that sending stuff into orbit is expensive and risky – risky, as you must send a delicate, and expensive, instrument into orbit on the back of a rocket. But the pay-off is huge, and we of course come to the (well justified) poster-child of astronomical imaging: the Hubble Space Telescope. The HST is relatively small in terms of its mirror's collecting area compared with the best ground-based telescopes (mirrors are heavy, and therefore expensive to get into space), but does not have to contend with either the absorption or distortion of light in the atmosphere, so it produces exquisitely sharp and sensitive images.

The second option is to figure out a way to correct those imperfections caused by the atmosphere using ground-based technology. That way, you can make use of the biggest mirrors that are simply too heavy to deploy in space. It *is* possible to beat seeing from the ground, and even to best the resolution

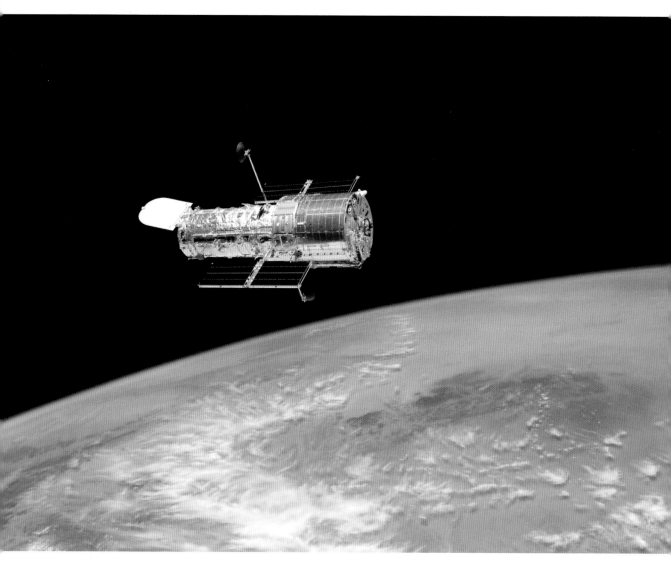

of HST. The trick is to actively control the telescope optics to compensate for the rapidly varying modulation of incoming light rays, in effect correcting the distortions imprinted by the atmosphere. This correction needs to be done several hundred times per second in order to work effectively. Sounds far-fetched? The technology exists, and is called adaptive optics. Here's how it works. Imagine dropping a stone in a still pool, and watching the ripples emanate away from the point of impact. Close to where the stone hit the water, the ripples are circular, but as we move further from the source and those circles grow larger they become more like sets of parallel waves. The

The Hubble Space Telescope, observing the universe from a vantage point above the Earth's atmosphere, has revolutionised our view of galaxies with its exquisite imaging capabilities. Launched in 1990, the HST continues to deliver outstanding science.

same principle applies to the light hitting the Earth's atmosphere from any distant astronomical source. By the time light from a distant galaxy arrives at Earth, the light is effectively coming in sets of perfectly parallel waves. When they pass through the atmosphere, this perfection is disturbed, distorting the parallel sets. This results in a blurred image. For science purposes we would like to take out this interference, returning the incoming waves to their parallel state.

The way to do this is to keep track of the distortions in a bright reference 'point' source, like a star. In the absence of atmospheric distortions, a star should appear as a single, steady point of light with a characteristic shape. If a bright star is not available near to where you are looking, then some telescopes are equipped with a powerful laser that can create a false star by exciting sodium atoms residing in a thin layer of the atmosphere, 100 kilometres up. By tracking the modulation of this reference source, the shape of the telescope mirror surface can be altered (by very small amounts) to compensate, bringing the image closer to what it would look like if the atmosphere wasn't there. One way to achieve this compensation is to deform the mirror using small pistons that can go up and down, changing the shape of the mirror surface to unwrinkle the incoming wave fronts. Think of it as analogous to throwing a bunch of tennis balls up in the air and then trying to catch them all at exactly the same time. When adaptive optics works, the results are spectacular, offering up to a 30-fold improvement in resolution compared with normal ground-based observations.

It's no good just capturing photons with a mirror. In order to do anything scientific, this energy must be recorded. Enter the charge coupled device, or CCD. The CCD has been the workhorse of virtually all astronomical detectors for over twenty years, and replaced the photographic plates of yore. Today the technology is ubiquitous in our lives. How does it work?

A CCD is a two-dimensional grid of detectors, analogous to pixels in a digital image (in fact, in their simplest application, they produce the content of pixels in a digital image). Each detector is made of a semi-conductor, typically silicon-based; a photon hitting one of these detectors can generate a small electrical charge. The amount of charge generated per photon that hits the detector increases linearly, so if we bathe the chip with many photons – in other words, we 'expose' the CCD – we can build up a large charge that corresponds to the amount of light that hit it during the exposure time. Charge can be manipulated with voltages, and so, after a suitable exposure

time, we can 'read out' the charge in each pixel by shuffling the signal in each detector to the edges of the CCD, where it can be amplified electronically and passed through a converter that turns an analogue voltage into a digital number (called an ADU). At that point we can save the information for posterity in a two-dimensional array: a digital image stored in memory. Then the fun begins.

For your digital camera, after the shutter has closed, that's the end of the story. The image you get on your screen is usually an extremely accurate record of whatever you took the photo of, and doesn't require much – if any – post-processing. But everyday photography enjoys the benefit of what astronomers crave, but generally lack: signal-to-noise. Simply put, the signal we are generally searching for (the light emitted by some distant galaxy) is often dwarfed by the emission from the sky, and can be comparable in size to the random fluctuations, or noise, in the read-out of every detec-

One of the four Very Large Telescopes of the European Southern Observatory in operation. The bright streak shooting into the heavens a powerful laser, used to create an artificial 'guide star' by exciting sodium atoms high in the atmosphere. The excited sodium glows like a star, the light from which can be used as a reference source to perform corrective 'adaptive optics', which can sharpen images taken from the ground by compensating for the blurring effect of the Earth's turbulent atmosphere.

tor. Sometimes, we even have to worry about the amount of 'dark' signal introduced due to the production of charge in each detector from the thermal production of electrons in the semi-conductor – present even when no light is shining on the CCD. In short, raw astronomical data is ugly. Not only do we usually have to combine many exposures of the same patch of sky to build up enough integration time to detect the signal we're after, but it takes a lot of post-processing to produce science-grade images, or even what we might call 'pretty pictures'. This is called data reduction, and is so named because we start off with a lot of data and whittle it down to a single image, rejecting a lot of it along the way.

CCDs are not the only detectors used in astronomy. We are constantly developing (or exploiting) technology to be able to detect other forms of radiation from distant galaxies. I'll give you an example. I'm writing this from a hotel room in Hilo, on the Big Island of Hawai'i. I'm here because I'm helping out with the commissioning of a brand new camera on the James Clerk Maxwell Telescope (JCMT) called SCUBA-2. SCUBA-2 is a camera sensitive to 'submillimetre' waves – to be specific, light with wavelengths of 450 and 850 microns. Here a traditional semiconducting device will not do; something more exotic is required. SCUBA-2 still makes use of a two-dimensional array of pixels, but this time each detector is a superconducting 'transition edge sensor' that is held at a temperature just above absolute zero. These devices can measure submillimetre photons by the small change in temperature they impart when they hit the detector, which changes the electrical resistance, which can be measured as a small change in voltage – typically a billionth of a volt. Voltages can be turned into a digital signal, which is then stored. Thus we have a way of recording the incident light. As these examples show, exactly how this is done in a practical sense varies as we move along the electromagnetic spectrum, but what we find in common is the conversion of incoming electromagnetic flux into a digital signal that can be calibrated in a way that tells us how much energy arrived at a certain frequency of light. This is the key for actually interpreting observations of distant galaxies.

Before SCUBA-2 starts its scientific duties, the behaviour of the camera and the data it produces have to be fully understood. Not only are we trying to use the instrument to find out something new, but we're trying to understand the instrument itself. It's been newly fitted to the telescope and as I write this it is in a stage of commissioning, involving many tests and tweaks. By the time you read this book, it will be conducting astronomical surveys for real.

The commissioning of any new instrument is exciting in itself (although frustrating and nerve-wracking for the technicians and engineers who built the thing). It's not a case of just plugging it into the back of the telescope and opening the shutter: in the case of SCUBA-2, first of all the whole thing must be cooled down to cryogenic temperatures, lower than one degree above absolute zero, then all of the individual detectors have to be tested: do they all work, do they all respond to incoming photons in the same way, what biases might there be? In addition, new software has to be developed to control the camera, and handle the raw data that comes off it. All these commissioning processes take time, but are vital for the successful execution of scientific experiments, because we need to understand exactly how the instrument works in order to interpret new results.

At submillimetre wavelengths, the bulk of the signal the camera sees is actually from the Earth's atmosphere, and this is extremely variable. The signal from the sky must be accurately removed, as must the random offsets and steps in gain, data spikes caused by various glitches and other gremlins. A prime example of one of the nuances of this new data is an unusual pattern noticed in some of the commissioning maps. Unfortunately, because the devices that read out the signal in the SCUBA-2 camera are also excellent magnetometers, we're also seeing some residual 'emission' in the maps caused by contamination from the Earth's magnetic field. Luckily, we can remove the signal using some clever signal-processing techniques, and by trying to shield as much of the field as possible from the sensitive instrument. The reason we need a submillimetre camera is, as we have seen, that galaxies emit a huge range of different forms of radiation, released from different components and physical processes going on within them. In the case of the submillimetre bands, this light is related to the cold dust and gas associated with star-forming regions. But we must be able to soak up *all* of the different forms of electromagnetic energy coming from galaxies.

We experience different aspects of the full gamut of electromagnetic radiation every day, whether we are being x-rayed in a hospital, using a microwave in the kitchen or tuning in an analogue radio. It's clear that the sources, and nature, of the radiation we come across every day are very different, and play different roles in our experience, but they are around us all the time – a sea of waves everywhere. We can only see the waves that our eyes are receptive to, but our radios and televisions can 'see' – in a sense – the photons with wavelengths much longer than visible light. Imagine if you could only see radio waves – the world would look very different. In fact, it would be unrecognizable compared with our everyday outlook. But the radio view would tell you something different about the world that your normal, visible-light view does not. Only in combining all the different views can we build up a holistic picture of how galaxies work. This is called the multi-wavelength approach.

No better example of this is presented than the multi-wavelength views of our own galaxy. The entire sky has been mapped by various telescopes, ranging from very high-energy gamma-rays and x-rays, through the ultraviolet and visible bands, to the near-, mid- and far-infrared and millimetre bands and finally the radio bands. Images of the whole sky at any wavelength are dominated by emission from the disc and bulge of our galaxy, and these maps are usually orientated such that the disc runs horizontally through the centre of the image, projected in what we call 'galactic coordinates'.

Optical, or visible, light shows the emission from the stars, but towards the mid-plane of the disc and in the central bulge there are dark patches where interstellar dust obscures the view. Turn to the near-infrared (a few microns in wavelength) and the view has changed. We still see the stars, but this time there are fewer dark patches; remember, near-infrared photons are not scattered and absorbed as easily as the optical wavelength photons, and this allows us to cut through that interstellar dust as if it wasn't there. We are now seeing light predominantly from the older stars in the galaxy, which emit most of their light in the near-infrared, with the bulge and disc glowing brightly. Turn to far-infrared emission, and we can see the glow of the interstellar dust itself, again concentrated in the disc, re-radiating the energy it has absorbed from the incident starlight. If we look at a very particular radio frequency, 1.4 GHz (equivalent to a wavelength of 21 cm), the atomic hydrogen in the galaxy is revealed. This time we don't see much of the bulge because most of the radio emission originates in a thin ribbon at the mid-plane, representing atomic gas within the dense disc of the Milky Way. We could go on; scan across the full electromagnetic spectrum and we get the whole shebang, and these different views represent different layers that we can peel away to help us understand the structure and physics of the galaxy. We can do this with our own galaxy, as well as other galaxies. Any single-wavelength view is incomplete; only when it is put together do we get the whole story.

Generally, when we simply take an image of the sky with a telescope and CCD, or some other detector, we're just collecting all of the light that is allowed to pass through some filter that is placed in front of the detector, or, in radio astronomy, the range of frequencies passed by some receiver, which work differently to CCDs. In the visible and near-infrared wavelength regime, the filters split the optical (visible) portion of the electromagnetic spectrum into chunks, moving from blue to red, and together are called 'photometric systems'. Each filter restricts the range of frequencies of light that can hit the detector. The widest of the filters – the ones that span the largest range in frequency – are called broad-band filters. Images of distant galaxies taken with broad-band filters are useful because they provide morphological information about the distribution of starlight: the shape of the galaxy (spiral or elliptical, say), the size of the bulge compared to the disc, and so on. Images like this are the pretty face of astronomy. But encoded within this broad-band light is far, far more information. The broad-band light can be split in exactly the same way as white light can be through a glass prism: the rainbow of colours that comprise the white light is separated because the monochromatic photons are refracted, or bent, by slightly different amounts depending on their frequency – their colour. So when white light passes through a prism, what comes out is a rainbow of colours; we have dispersed the light.

Imagine holding a prism up to sunlight, and projecting the rainbow onto a screen. If you were to measure the intensity of the light in each of the colours, you would find that the intensity rises and falls with a particular shape, peaking around the green/yellow mark. This is the 'spectrum' of our Sun; simply the distribution of energy emitted as a function of frequency. We can use the spectrum to learn about the composition and physics of the Sun. The Sun is just one star; when we measure the spectra of whole galaxies, we're seeing the combined light from billions of stars as well as the gas among and between them.

To actually measure astronomical spectra in practice, we can still use CCD detectors to record the photons, but a crucial extra piece of hardware is required – a dispersive element. This dispersive element can be a prism, or, more commonly today, a 'grating' (effectively a set of narrow slits aligned closely together, which disperse light due to diffraction as waves pass through the grate) or a 'grism', which combines principles from the grating and the prism. Whatever the dispersive element, the key is to split the light according to its component frequencies, so white light, or whatever range of frequencies have been passed by a filter, becomes a rainbow. This comes at a cost in terms of observing time, because when we disperse it, the total energy in a beam of light is spread out according to the intensity distribution of the spectrum. Like a knob of butter spread over toast, the light is smeared out over a larger number of CCD pixels than would be the case if the light was not passed through a dispersion element. So for a given target, typically a much longer exposure time is required to obtain a spectrum compared with simply taking an image of the thing, where all of the light is concentrated onto a smaller number of pixels.

Sensitive instrumentation and large telescopes are not enough; the physical location of these facilities is also of great importance, and as we push the frontier of research, astronomers have demanded ever higher standards in the quality of the observing sites where we choose to put our 'scopes. Cerro Paranal is a modest, 2,000-metre-high mountain in the Chilean Atacama Desert, about 120 kilometres south and inland from the northern coastal town of Antofagasta and just over 1,000 kilometres north of Santiago. The reasonably high altitude, exceptionally arid conditions, stable atmosphere and remoteness make this an exquisite location for astronomical observations. From the southern hemisphere it's possible to see the Magellanic Clouds

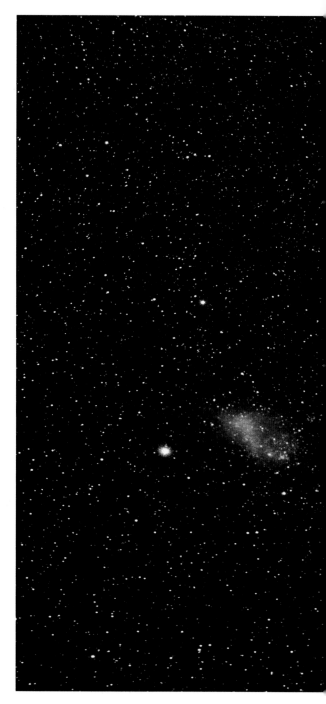

(Large and Small, or LMC and SMC). These are two 'dwarf' galaxies – low mass, relatively faint and irregular looking – that are satellites to the much larger Milky Way, and many famous constellations, like the Southern Cross, that are not visible from the northern hemisphere. Many of the interesting galaxies that we want to observe, and indeed a large fraction of the sky in general, are only visible from the southern hemisphere, just as some are only visible from the northern; this is why we require telescope facilities in both hemispheres – another limitation of being observers trapped on the surface of a small sphere.

Paranal is the home of the European Southern Observatory's formidable Very Large Telescopes (VLTs). The VLTs are four telescopes in the '8 metre' class, which means that their light-collecting primary mirrors are roughly 8 metres in diameter. Only the two Keck telescopes atop Mauna Kea on the Big Island of Hawai'i (one of the other premier sites for ground-based astronomy), with their 10-metre segmented mirrors, currently exceed the 8 metre class of optical telescope. We do have telescopes on Earth with dishes much larger than 10 metres, but these are designed to detect photons with longer wavelengths, like radio waves. The reflecting surfaces of radio telescopes are not made from silvered glass because radio waves are easily reflected by other types of material, like concrete or aluminium. Making very large collecting dishes from these materials is far easier than from glass, which has limited the physical sizes of telescopes that detect visible and near-infrared photons.

The mirrors – or 'light buckets', as we call them – are just part of a telescope. The VLTs are equipped with an arsenal of equipment to capture, record and measure the photons that are collected: cameras, spectrographs and integral field units, capable of detecting photons from ultraviolet wavelengths (before the Earth's atmosphere starts blocking everything out below a wavelength of about 300 nanometres) to the near-infrared,

The Magellanic Clouds, two 'satellite' dwarf galaxies of the Milky Way. The Clouds are named for Ferdinand Magellan, the Portuguese explorer whose voyages took him to the southern latitudes where these cloud-like patches of light are visible, although the structures were noted by other late medieval European explorers in the fifteenth century, and earlier by Persian astronomers in the tenth century. Of course, the Clouds would also have been well known to indigenous peoples of the southern hemisphere as familiar sights in the night sky for thousands of years.

at a wavelength of about 2 microns. Unless a telescope or instrument attached to it is named after a person (usually a famous astronomer), it generally becomes known by an acronym (actually, even the Hubble Space Telescope, named for the astronomer Edwin Hubble, is generally simply referred to as HST). Thus we have the ESO VLTs, equipped with such instruments as ISAAC (Infrared Spectrometer And Array Camera), FLAMES (Fibre Large Array Multi-Element Spectrograph), HAWK-I (High Acuity Wide field K-band Imager), and VIMOS (Visible Multi Object Spectrograph), among others.

The precise detail of these instruments is not important; I've just picked them at random from a long list of cameras and detectors currently in use. The point is that different instruments can be used to answer different science questions. For example, we might take a simple image of a galaxy with the HAWK-I camera, which has a CCD sensitive to near-infrared light. Perhaps we want to map out the older stars in a particular galaxy. Once we have our image, we could use ISAAC to measure the spectrum of that galaxy's near-infrared light. This is done by placing a narrow slit in the path of the light coming from the galaxy (and therefore isolating it from all the other light coming from the sky, or indeed from other parts of the galaxy itself) and then dispersing the near-infrared light from the galaxy, breaking it down into its components in the same way a rainbow forms from sunlight through raindrops. This tells us how much energy is being emitted at different frequencies, and the detailed shape of the spectrum can provide information about the composition of the stars and gas, and also the relative motions of stars and gas within the galaxy.

The telescope superstructure – the protective dome, mirrors, struts, control room housing computational infrastructure and human operators, and even the concrete plinth that the whole thing sits on – is fairly permanent, but the great thing about the *instruments* is that they can be exchanged with new ones if and when old ones break or become obsolete. Old instruments can be removed and new ones slotted into special ports where light can be directed after it has been collected by the primary mirror; the light is diverted this way and that by way of a carefully engineered optical path. A great deal of technological effort and ingenuity goes into devising new instruments for telescopes to meet the demands of science; these demands are usually for higher sensitivity and efficiency, both achieved as cheaply as possible, and this drives forward technology outside the realm of astronomy, in areas like CCD-detector technology and optics. There are also synergies with quite disparate fields, such as medicine: principles of wave-front sensing used in adaptive optics can be applied to compensate for macular degeneration in the human eye, for example.

The astronomer's residence, or *residencia*, at Paranal is really quite extraordinary. The residencia houses visiting astronomers and other staff working at the site, and can best be described as a Bond villain's lair (it was actually used as a location in the Bond film *Quantum of Solace*). The *residencia* is located a few kilometres from the telescopes, which perch atop Paranal itself. The geometric facade of concrete, coloured the same red, Martian hue as the desert, and glass is the subtle outward sign of the building, but its airy guts are dug into the side of the mountain, and contain dorm rooms, a restaurant, gym and offices. The entrance to the *residencia* is a cavernous atrium filled with subtropical plants and a swimming pool, providing some evaporating moisture to dampen the otherwise extraordinarily bone-dry air, and also to cool off hot astronomers. It is an incredible place to work.

On an observing trip to the VLT, I had the privilege of witnessing the construction of a new telescope – the Visible and Infrared Survey Telescope, or VISTA. VISTA is a 4-metre class telescope, half the size of the VLTs, and is designed for a single purpose: big surveys. It is equipped with a huge, 67 megapixel camera called VIRCAM, which can snap an image of the sky covering a diameter of 1.65 degrees (around three full Moon diameters) at once. This large area, coupled with the excellent sensitivity of the camera, makes VISTA an efficient tool for making big, deep maps of the sky. This is useful, because with large surveys we can efficiently detect many thousands of distant galaxies at once (which is important for statistical studies), and learn about their distribution in space. VISTA is conducting several wide-field extragalactic surveys (it is particularly good at detecting extremely distant galaxies), as well as surveys of our own galaxy, of course.

Why go to all this effort? Why drag many millions of dollars' worth of telescopes and equipment and people to some of the most remote and inhospitable places on Earth just to collect a few of the photons that gently shine down on us from all the luminous products of 14 billion years of cosmic history? When you sit down and work it out, the amount of energy we actually collect from astronomical sources is fantastically small: the amount of energy received from a galaxy I am studying, per second, per unit area, is about a thousand times smaller than the kinetic energy of a single snowflake hitting a detector the size of the United Kingdom. The galaxies play an insignificant role in our lives. We know that the Earth is round, that it is not the centre of our solar system, and that there are other planets. Do we really *need* to know what's out there, beyond the Milky Way, where we will never voyage? I would, of course, argue yes.

It's true that the things we are learning about the universe now don't have the same direct practical implications as, say, the knowledge that the Earth is a sphere. One day, humans, or more likely our species' descendants (or autonomous exploring machines they may build), will explore the wider galaxy, but this will probably not be for hundreds or thousands of years. And it is almost certain that we will *never* visit an external galaxy (there's more than enough to keep any civilization occupied in the Milky Way for aeons, anyway). Closer to home, human and robot exploration of the solar system will certainly be vital for future generations, either by the colonization of the Moon and Mars, or in mining asteroids for resources. But why bother delving into the physical details of the extragalactic universe?

Mankind is driven by an atavistic fascination with the natural world and a desire to understand its workings on the most complete level. That requires an understanding of the universe as a whole: its contents and its evolution. When I look up at the stars I'm not satisfied with a superficial appreciation of their delicate beauty and mystery. I want to know *what* a star is. What is a star made of, how are stars made, how far away are they? Not knowing the answers to these questions, for me at least, is like watching a thunderstorm and not knowing what a raindrop is. Astronomy provides the answers to some of our more fundamental questions; it gives us a clearer picture of nature and our place within it. While the questions we want to answer are simply motivated by a desire to understand, the scientific process itself has proven time and again to lead to new, practical solutions to 'real world' problems. A good example is in the development of WiFi, the wireless system that allows computers and other devices to communicate through the air. The algorithms that ensure the radio signals are transmitted and received cleanly (that is, with little interference) were the result of the development of signal-processing techniques in radio astronomy. But I think that what it comes down to is this: we haven't stopped doing astronomy for the same reason we haven't stopped making art – it is just part of us.

What we know

Astronomy is an ancient science; the first steps we took were small and slow, but now we are at a sprint. Astronomy has never stopped moving forward. In fact, extragalactic astronomy is still relatively young as a discipline, and we are learning new things at an astonishing rate. We have only really come to understand that our own galaxy is self-contained, and that there are many other galaxies external to our own, over the last few generations. In contrast, we have known that we live in a heliocentric solar system for nearly half a millennium. Let's start with a bit of historical background on how the field of galaxy evolution came to be.

In the geocentric and early heliocentric models of the universe, which placed the Earth and the Sun, respectively, at the centre of the universe, the stars were not really considered to be three-dimensional entities distributed in space. Rather, they were modelled as 'fixed' on the celestial sphere, just beyond the planets. In the late eighteenth century, the Herschels (siblings William and Caroline) searched for evidence of a regular distribution in the stars by counting the number of stars in different patches of sky. They found that the counts decreased away from the Sun, and concluded that the Sun was therefore the centre of the universe. However, the Herschels' observations did not take into account that pesky, obscuring interstellar dust that is particularly prevalent in the direction of (what we now know to be) the centre of the galaxy, which masked the true number of stars. This is an example of a situation where contemporary instrumentation and observational techniques were inadequate for answering the major questions of the day.

The first, remarkably prescient, proposal that we live in a distinct galaxy of which the Sun and Earth are just a minor component came from the English astronomer Thomas Wright of Durham, who in 1750 published *An original theory or new hypothesis of the Universe*. In this he argued that the band on the sky called the Milky Way appears for the reason that we live in a flattened disc of stars. He even proposed, or at least speculated, that the 'cloudy spots' – some of the nebulae in the sky – were external systems, far away from the disc. This idea was taken forward by the philosopher Immanuel Kant a few years later. Kant is often cited, ahead of Wright, for the idea of the 'island universe', Kant's term for the hypothesis that the 'spiral nebulae' were other, distant galaxies.

Even into the 1920s, debate raged about the true nature of spiral nebulae and the size of the universe. The 'Great Debate' between Harlow Shapley and Heber Curtis in 1920 illustrates this. Shapley argued that the universe was basically defined by the Milky Way, with space pervaded with stars and gas and dust. The spiral nebulae, in Shapley's view, were entities within this all-encompassing stellar system. Curtis, on the other hand, argued for the island universe model, where the cosmos is a vast space, with galaxies forming discrete concentrations of stars, separated by vast distance. In fact, it was the extremely large distances implied between the Milky Way and the other spiral nebulae that made opponents of the island universe model sceptical.

In the end, the island universe model has been proven correct. The picture we now have is that, indeed, our galaxy is not at the centre of the universe; it is just one of billions, with each galaxy separated by distances much larger than the size of the galaxies themselves. So how did we figure this out empirically?

The main evidence came later in the 1920s, when astronomers studied a particular type of star in the 'spiral nebula' in the constellation of Andromeda. This object is also known as M31, as it was the 31st entry in eighteenth-century French astronomer Charles Messier's catalogue of celestial objects of interest. On a dark, clear night you should be able to see M31 (or Andromeda, as it is colloquially known) with a pair of binoculars, or even with the naked eye, as a faint, elongated smudge of light. The stars in question are called Cepheid variables, and are peculiar in that, unlike most stars, they pulse in brightness – roughly doubling over the course of a regular cycle. Cepheids are named after the star Delta Cephei, the fourth brightest star (that's where the Delta comes from) in the constellation Cepheus, and one of the first of its kind to be discovered in the eighteenth century.

Cepheids pulse in brightness because the star itself is physically expanding and contracting. The opacity of gas in the photosphere of the star (the layers of gas on the outside) determines how much of the light produced by nuclear fusion in the core can actually escape from the star, as opposed to bouncing around among the gas through a process of absorption and re-emission. The opacity of the photosphere is related to the pressure of the gas; during the cycle of expansion and contraction there is a regular change in gas density, the pressure changes, and therefore the total number of photons that are emitted varies accordingly. What we see is a regular brightening and dimming of Cepheid light.

The Great Galaxy in Andromeda, seen here in a wide-field view that is dominated by stars within our own galaxy. With the human eye Andromeda looks like a faint, fuzzy patch of light amongst the stars, and even with the early telescopic views it was thought that this 'spiral nebula' and others like it were part of the Milky Way. After all, the Milky Way does contain nebulous regions, like the Orion nebula, and other exotic objects, like globular clusters, so why should Andromeda be any different? However, when the distance to Andromeda and the other nearby galaxies was determined from Cepheid variable star observations, it became clear that they were external systems separated from us by great tracts of space. This cloudy spot is about a million times more distant than the stars in this image.

The typical length of a Cepheid pulsation cycle is extremely short in astronomical terms. In fact, the variation is very much on a human timescale, ranging from a few days to several weeks. If you're so inclined, you can even conduct experiments from your back garden with just a small telescope, measuring the brightness of known Cepheids from night to night, tracking their light beats. Perhaps the easiest Cepheid to spot (in the northern hemisphere) is the pole star, Polaris.

Now, it turns out that the Cepheid cycle has a very useful correlation: there is a tight relationship between the length of an individual star's pulsation cycle (the time between bright peaks) and the average luminosity of the star. Cepheids with longer periods are more luminous than those with shorter periods. This discovery was made by the American astronomer Henrietta Leavitt, who published her observations of Cepheids (that reside in the Large Magellanic Cloud) in 1912.

Why is the period-luminosity relation useful? Because if we know the intrinsic luminosity of an object (the total amount of energy it emits every second), we can compare this to its apparent brightness on the sky (the flux we measure with a telescope) and thus work out how far away it is. As we discussed earlier in this chapter, the observed brightness of a source drops according to a well understood 'inverse square law', so if you have an independent measurement of the intrinsic luminosity – the total amount of energy being released – you just put those numbers into the inverse square law and work out the distance. Actually, around the same time as Leavitt's discovery, Ejnar Hertzsprung, a Danish astronomer, calibrated the period-luminosity relation using distances to Cepheid variables in the Milky Way for which he had measured parallax, thus tying the Cepheid technique to an independent distance-measurement technique. Accurately measuring physical distances is one of the hardest problems in astronomy, and so we call objects like Cepheids 'standard candles', because they are objects for which the luminosity is well calibrated.

Edwin Hubble and Milton Humason found that the Cepheids in M31 are so distant that they must lie far outside our Milky Way. The discovery of these distant Cepheids pretty much settled the island universe debate. M31 was certainly external to the Milky Way by a long margin. When imaged properly so one can capture the faint emission of the extended stellar disc, in terms of area on the sky, M31 is larger than the full Moon, but it is actually around a million times further away than the nearest star. If the starry disc of the Milky Way was contained within the M25 around London, Andromeda would be located somewhere near Moscow. The field of extragalactic astronomy, or rather extragalactic galaxy studies, had begun. When we look at the deepest optical images of M31 now, given all we know about external galaxies, it almost seems obvious that this nebula is a self-contained and distant star system, but this was far from clear at the time, and it is important not to underestimate how important this leap was in our understanding of the universe. Just like all theories and models of the universe, then and now, we aim to empirically test, corroborate and refute, no matter what gut instinct tells us.

As astronomers began to explore more local galaxies – the ones that are close enough to the Milky Way, and thus bright enough to be detected with early twentieth-century telescopes – something even more remarkable was discovered. First, it appeared that the light from distant galaxies was redder than expected. I'm not talking about a vague difference in shade; rather, all of the light emitted by a distant galaxy appeared to be systematically shifted to longer – that is, redder – wavelengths. The clearest signature of this effect is seen in the spectra of galaxies, which are the astronomical equivalent of fingerprints.

The power of spectroscopy

A spectrum is simply a measurement of the amount of energy output by a luminous object, be it a candle flame or galaxy, at different wavelengths (or equivalently, frequencies) of light. For example, if we take the light from the Sun and split it through a prism, what we find is a characteristic 'continuum' of light – the rainbow – with an intensity that peaks at a wavelength of about 500 nanometres, corresponding to yellowish light. The Sun does emit radiation that is not in the human-visible part of the spectrum, like the ultraviolet and infrared, but the emission is weaker here. The spectrum is not perfectly smooth either. The bright continuum emission is indented with thousands of dark patches at certain wavelengths: these are the absorption lines, and they are caused by particular elements within the Sun that absorb photons of very specific energy (and therefore very specific frequencies). These dark lines are called 'Fraunhofer lines' after Joseph von Fraunhofer, a nineteenth-century German optician.

Fraunhofer was a master of his craft, influential in developing the spectroscope and a pioneer of the field of astronomical spectroscopy; the solar absorption lines are named in his honour. Under certain conditions, some elements can also emit, rather than absorb, photons of a particular energy. These are called emission lines, and appear as bright spots, or spikes in a spectrum. If you have ever spilled some salt into a flame, then you will have noticed that the flame suddenly burns a bright yellow; this is because, when the salt is broken down, the sodium in it is ionized because the energy of the flame is enough to remove an electron from around the nucleus of the sodium atom. When an electron recombines with the atom (or, more likely, another atom that has also lost an electron), the energy that went into removing it is released. Since this is a very specific energy change (quantum mechanics tells us that the different possible energy levels in atoms are discrete) it corresponds to a very specific colour. In the case of sodium, the wavelength of light released is exactly 589.3 nanometres. This is what gives sodium-based street

A clearer view of the Andromeda galaxy, also known as Messier 31 (M31), in ultraviolet light from the GALEX satellite. In this image the complex structure of the galaxy is evident, with the spiral arms in the disc encircling a central hub. M31 is not unlike the Milky Way. Telescopes that are sensitive to ultraviolet photons can reveal the emission of young, massive stars that are prevalent in the gas-rich discs of spiral galaxies where new stars are being formed, so the spiral arms are prominent in this image. Ultraviolet light cannot pass through the Earth's atmosphere, and so these observations must be conducted from space.

lights that characteristic yellow colour. If you could take a spectrum of a sodium street light, then you would see that most of the light is emitted in one of these emission line spikes. So, not only can we use the shape of the spectrum to tell us something about a star, or galaxy, but the emission and absorption lines tell us about their chemical composition.

From laboratory tests here on Earth, and from atomic theory, we know the precise wavelengths of the emission and absorption lines produced by all the different elements. These can be 'matched up' to the emission and absorption lines observed in stars and gas locally and in distant galaxies. When we measure the spectra of distant galaxies, we find that all the spectral features are systematically shifted along the scale in wavelength, but the relative distance between individual emission and absorption lines within the spectrum is the same as would be measured here on Earth.

For example, a common emission line to be found in galaxies is called Hydrogen-alpha (H-alpha), which is one of the emission lines emitted by the ionized gas near new stars that we discussed in chapter One. H-alpha is the principle line in the 'Balmer' series of hydrogen emission lines, which goes H-alpha, -beta, -gamma and so on. To recap, when the hydrogen atom is hit by a photon with the right energy, an electron can escape from its orbit around the nucleus (we say the atom is ionized). When the electron recombines and settles back in its original energy level, a photon is released. H-alpha light, when measured on Earth, has a wavelength of about 650 nanometres, but we might measure H-alpha in some distant galaxy and find it to be closer to 2 microns in wavelength. We know it is H-alpha and not some other line because of its position relative to other emission lines and spectral features, which serve as a kind of barcode identification. What's going on? It's not that the underlying physics governing how those photons are being emitted is varying from galaxy to galaxy.

This effect is called the redshift. Redshift can be thought of as the light equivalent to the change in the pitch of a siren on a police car that zooms past you (this is called the Doppler effect). If you were in the police car, you would not hear this change of pitch, because you are in the same 'frame of reference' as the siren. The same thing applies here. If we were to visit that distant galaxy so that we were in its frame of reference, or equivalently if we were not moving relative to that galaxy, we would measure the H-alpha line at the 'rest frame' wavelength – the same wavelength we measure in the laboratory here on Earth.

But what if we are not in that galaxy's 'rest frame'? From our point of view – from *our* frame of reference – if a distant galaxy is hurtling away from us, then, like a police siren Dopplering away, we measure the light emitted by that galaxy as shifted systematically to longer wavelengths. The overall shape of the spectrum of that galaxy does not change because all of the gas, stars and dust are moving roughly in tandem; everything just gets redder to us. Of course, if the source was moving toward us, then the light would be shifted to shorter wavelengths – it would be blueshifted. The redshift is measured through the ratio of the observed and 'rest frame' wavelengths (or frequencies) of light. So, a redshift can be related to a galaxy's velocity, relative to the Earth, along our line of sight.

Now, here comes the seminal moment that really marked the beginning of the age of extragalactic astronomy, and, for that matter, what we now call 'observational cosmology'. Edwin Hubble, working at the famous Mount Wilson Observatory, took the redshifts of a sample of galaxies that were previously measured by the often overlooked astronomer Vesto Slipher. Hubble and Humason had compiled distances from Cepheid variable observations of these galaxies that Slipher had measured redshifts for, and when the redshifts and distances were compared, a correlation was found: in general, more distant galaxies had larger redshifts. In fact most of the external galaxies had positive redshifts, with only a few showing blueshifts. In 1929 Hubble published the discovery.

It's important to note that there were several other astronomers involved in early theoretical work. For example, earlier in the decade, Alexander Friedmann and Georges Lemaître, working independently, derived forms of what would become known as Hubble's law using Einstein's general theory of relativity. There were also other observers conducting independent observations that were beginning to reveal this picture of an expanding universe. Thus there is some controversy over who exactly should get credit, although usually Hubble alone is cited as the discoverer.

Whatever the politics of the discovery, the implications of this experimental evidence were profound. It showed that not only is the universe filled with galaxies separated by vast distances, but this combination of data – the Cepheid distances and the redshifts – implied that galaxies are generally moving away from each other, and those that are further away appear to be moving faster. The conclusion was clear: the universe is *expanding*. This was, and with the continual addition of more data remains, some of the

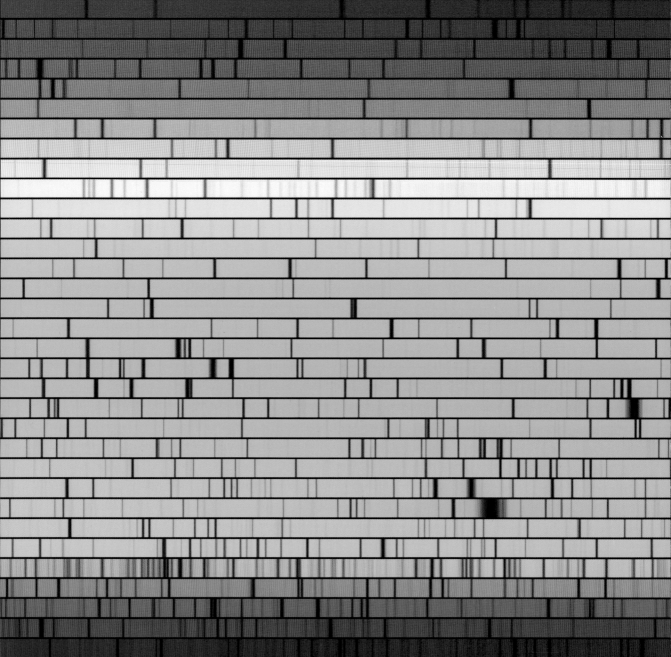

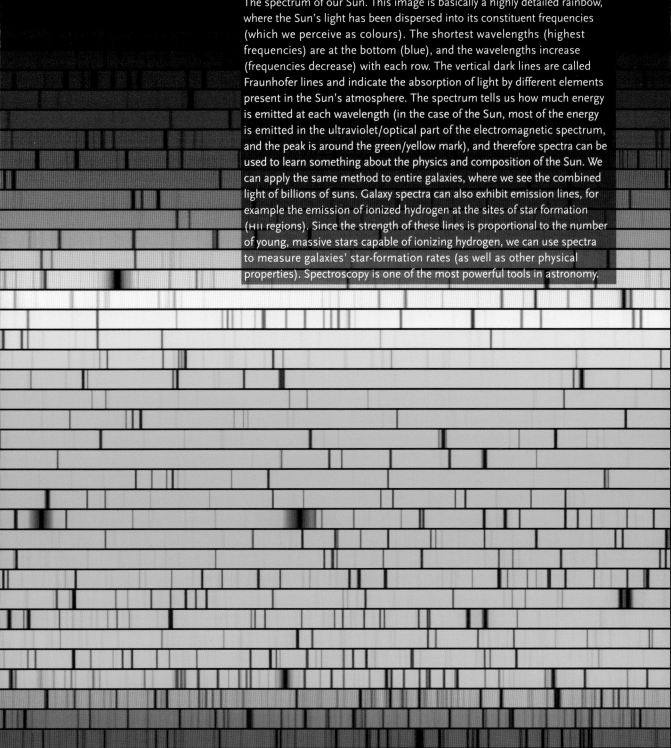

The spectrum of our Sun. This image is basically a highly detailed rainbow, where the Sun's light has been dispersed into its constituent frequencies (which we perceive as colours). The shortest wavelengths (highest frequencies) are at the bottom (blue), and the wavelengths increase (frequencies decrease) with each row. The vertical dark lines are called Fraunhofer lines and indicate the absorption of light by different elements present in the Sun's atmosphere. The spectrum tells us how much energy is emitted at each wavelength (in the case of the Sun, most of the energy is emitted in the ultraviolet/optical part of the electromagnetic spectrum, and the peak is around the green/yellow mark), and therefore spectra can be used to learn something about the physics and composition of the Sun. We can apply the same method to entire galaxies, where we see the combined light of billions of suns. Galaxy spectra can also exhibit emission lines, for example the emission of ionized hydrogen at the sites of star formation (HII regions). Since the strength of these lines is proportional to the number of young, massive stars capable of ionizing hydrogen, we can use spectra to measure galaxies' star-formation rates (as well as other physical properties). Spectroscopy is one of the most powerful tools in astronomy.

most compelling evidence for the origin of the universe in a hot Big Bang. Simply turn the clock backwards: things that are moving away from each other must have once been closer together. Run the clock back far enough and you get to a point where all matter and energy were condensed into a volume far, far smaller than their distribution today. Some mechanism, which we call the 'Big Bang' (actually, this started out as a derogatory term for the theory), caused an explosive expansion from a single point, which we assume was the starting point of our physical universe. Whether anything came before this is a matter of speculation and controversy, in part because it is difficult to test empirically.

The precise nature and mechanism of the initial expansion in the first few moments of the universe, and its continuing expansion today, are what we might term 'cosmological questions', and we're not going to focus too heavily on those here. We're interested in the galaxies themselves, caught up with this cosmic flow, and how they formed and evolved within the universe that sprang out of the hot conditions of the Big Bang.

Back to the spectra. Our ability to measure the spectra of galaxies is an essential part of our toolbox. Redshift can be used as a tool to map out the galaxy distribution in some sort of three-dimensional context, because we know that galaxies with larger redshifts are further away. But spectra have further applications – as we have seen, they contain important information about the internal contents, chemistry and dynamics of distant galaxies.

The Sun's spectrum is complex; its detailed shape basically encodes information about the chemistry of the star and how much energy it emits. We can measure the Sun's spectrum very well because it is so bright. But the Sun is just one star. When we take the spectrum of an entire galaxy, what we actually measure is the superposition of the light from billions of stars of different ages, masses and metallicities. Not only that, but we get all of the interstellar material too – gas and dust between the stars. If all the stars were of the same mass and age as the Sun, and there was no interstellar material, the spectrum of a distant galaxy would be pretty much the same shape as the spectrum of the Sun. But there is a range of stellar types in whole galaxies; not all stars are like the Sun. This results in differences in the shape of the spectral continuum from galaxy to galaxy, and we can use these differences to classify galaxies of different types.

Galaxies that are actively forming many new stars produce lots of emission in the ultraviolet and blue part of the spectrum, because this is the light produced by the very massive but short-lived stars. In other words, if we see a galaxy with lots of ultraviolet emission, then we know that it must contain many young (typically very massive) stars, and therefore the galaxy must be actively forming new stars, since these massive stars don't live for very long (in the order of millions of years). The ultraviolet luminosity can therefore be calibrated into a star-formation rate. The ultraviolet light produced by these new stars has another effect on the spectrum: it can ionize the interstellar hydrogen in the vicinity of the stellar birth grounds, creating the HII regions we talked about in chapter One. This produces strong emission lines in the spectrum (mainly lines of hydrogen and oxygen in the visible part of the spectrum), and the presence of these emission lines is another classification and calibration tool. Again, the strength of the observed emission line can be turned into a star-formation rate because we have a good idea of the number of ionizing photons per star required to produce it.

A more unusual view of galaxies. This image shows multiple spectra of several distant galaxies observed with the VIMOS multi-object spectrograph, an instrument on the Very Large Telescope. Spectroscopy disperses light according to frequency, like a rainbow, allowing us to examine galaxies' emissions in great detail, learning information about their motion and chemical make-up. In this image each vertical strip represents the spectrum of a single galaxy, and the bright horizontal lines are from emission features in our own atmosphere. The much fainter vertical lines seen in some of the strips are the emissions of the galaxies themselves.

Galaxies that are not forming new stars, but instead have a very mature, old stellar population, do not produce much ultraviolet light or gas emission lines. Most of the energy comes out in the redder, longer wavelengths of the visible and near-infrared part of the spectrum. These galaxies also have strong absorption lines, produced by metals that have accumulated from the process of stellar evolution over the lifetime of the galaxy. Prominent absorption lines in these galaxies come from the elements calcium and magnesium (again, in the visible part of the spectrum).

So spectra can be used to learn something about the internal conditions and average ages of other galaxies, and can be used to classify them into different types based on the features we see. Care must be taken, however. For example, it is possible that actively star-forming galaxies might *not* have much ultraviolet emission or exhibit particularly strong emission lines. One could therefore conclude that the level of star formation is low. The catch is that some galaxies contain copious amounts of interstellar dust – those silicon and carbon grains – which often surround the star-forming regions. As we know, dust absorbs ultraviolet and

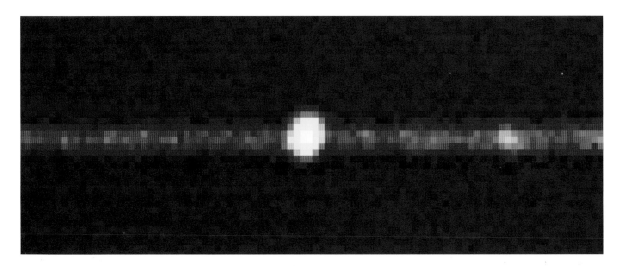

optical photons, and so 'reddens' the spectrum, suppressing the blue light coming from the new stars and the emission lines they produce when the natal gas clouds are irradiated. Unfortunately, the dust is often thickest – we say the optical depth is highest – right around the sites of formation of the new stars.

In some cases the reddening is so severe that it results in serious under-estimates of the star-formation rate of a galaxy. One way to get round this issue is to measure the amount of infrared light being emitted by the galaxy. As the dust absorbs ultraviolet photons, it heats up – typically to a temperature of between a few tens of degrees and a hundred degrees above absolute zero (depending on where the dust is, relative to the stars). This sounds cold, but in fact *any* object that has a temperature above absolute zero (-273 degrees Celcius) releases thermal energy. You emit infrared radiation at a wavelength of about ten microns. Cooler objects emit infrared radiation at longer wavelengths and vice versa. In the case of interstellar dust, the peak of the thermal emission is at about 100 microns (but with a broad spread). This is one way in which we can detect star formation that is hidden by dust: by searching for the telltale infrared emission caused by the heating of obscuring interstellar dust by starlight.

Spectrum of a distant galaxy showing a bright emission line representing ionized oxygen; the brightness of this emission line can be converted to a star formation rate for the galaxy. This spectrum was obtained with the Very Large Telescope FORS instrument.

Mapping the universe

Spectroscopy allows us to in some sense 'classify' galaxies, as well as helping us to place them in some three-dimensional context, thanks to redshift. But how are they actually distributed in space? Let's explore our 'local' volume in some more detail. Imagine that we could cut out a cube-shaped part of the universe and examine all of its contents in detail. Let's define a cube that is centred on the Milky Way, where each side is 20 million parsecs (twenty megaparsecs) long. This is a reasonably large chunk, even in cosmological terms; it contains a good sampling of the local universe. What does this cube contain? To make our visualization a bit easier, let's shrink this box down so that it is just 1 metre on a side; small enough to fit into a room. Imagine this cube of space in front of you, like a three-dimensional model (at this point, it would be great if holographic projectors existed, but for now just close your eyes and imagine. If it helps, find a big cardboard box).

In this scaled-down model, the size of the Milky Way itself, right in the centre of the box, would be just a millimetre across – barely visible to the eye. The tiny Milky Way is surrounded by a few of our dwarf galaxy companions, the Magellanic Clouds, as well as several other satellite systems, all within a few millimetres at this scale. The distance to M31, our nearest neighbour of similar type, is about 4 centimetres. Within a radius of 10 to 15 centimetres from the Milky Way, there are about 50 to 60 other galaxies. This is called the Local Group, and represents our cosmological back yard.

About 20 centimetres away in the direction of the constellation Centaurus (if we were sitting in the Milky Way looking out) is another grouping of galaxies, surrounding a large elliptical galaxy called Centaurus A, or Cen A for short. Cen A is actually a powerful radio galaxy. When viewed in the radio part of the spectrum, it reveals two large 'jets' of radio emission extending out from the centre of the galaxy, well exceeding the distribution of stars. Cen A is another reminder that we need multiwavelength views to get the full picture. The formation of these radio jets is related to what lies at the centre of the galaxy – a supermassive black hole (which we'll discuss later). The grouping of galaxies around Cen A is called the Cen A subgroup; often we find that galaxies are clustered together around the most massive galaxies in the universe, among which Cen A can certainly be classed. There are other groupings too – there is a large spiral galaxy in the direction of the constellation Hydra called Messier 83 (also an entry in Charles Messier's catalogue),

or the Southern Pinwheel galaxy. It is a beautiful spiral galaxy orientated so that we can see its face (our Milky Way would look similar, viewed from above the disc). In our box model, M83 is about 23 centimetres from the Milky Way. Like our Local group, M83 is also surrounded by a small group of galaxies, called the M83 subgroup. Many galaxies tend to be congregated in small groups, and we often find huge tracts of space with very few or no galaxies at all. These are called voids. Then we also find huge congregations of galaxies, called clusters.

Beyond the edge of our box, 80 centimetres away from the centre of our model, is a vast congregation of thousands of galaxies packed within a sphere about 20 centimetres across. At its core are several very large galaxies that don't look like the Milky Way, M31 or M83; these are not flat discs, but bulbous, symmetric elliptical galaxies, not dissimilar to Cen A. This is the Virgo cluster (so-called because from the vantage point of the Earth, the cluster lies in the direction of the constellation Virgo). Clusters are vast swarms of galaxies bound together by gravity, and the most massive entities in the universe. For reasons we'll explore in more detail in the next chapter, the properties of galaxies in high-density regions such as these clusters are different from those in the average 'field'.

Another view of Centaurus A, this time including submillimetre (orange, tracing cold gas and dust) and x-ray (blue, tracing very hot gas) light. Now we see the two jets of emission emerging from the galaxy. Cen A is a powerful radio galaxy (it happens to be one of the closest radio galaxies to the Milky Way), containing an active galactic nucleus that is responsible for this emission. This is an excellent example of why a multi-wavelength view of galaxies is required: we need to capture *all* the different emission features if we are to understand their nature.

This description of our local volume was not meant to be complete, but just gives a flavour of the distribution of galaxies in the universe, and the relative scales involved. You'll notice that most of the box is just empty space – the diameter of our own galaxy is just a tenth of 1 per cent of the size of that box. The other galaxies we met, although they come in a range of physical sizes (the ellipticals being the biggest), only occupy a tiny fraction of the total volume of space. You'll have noticed that the galaxies are not randomly distributed in space – they tend to be arranged in groups and clusters, and if we looked at all of the galaxies, we'd find that these groups and clusters all link up with a kind of galactic filamentary structure. The formation of these structures is due to gravity, and the formation and evolution of galaxies within this structure – how galaxy properties change as a function of their location within the 'large scale structure' – is an active area of research, and actually a large part of my own research.

We know the contents of our local box so well because of a continuing effort to map out the locations and properties of galaxies in the universe.

The Coma cluster, the most massive structure in the local universe, where thousands of galaxies congregate at high density. Clusters represent the parts of the universe that were the largest density fluctuations in the matter field shortly after the Big Bang. Under the constant attention of gravity, these perturbations have grown and accreted matter over time, evolving into mammoth structures like this. Clusters are inhabited by some of the oldest, most massive galaxies in the universe (ellipticals) and can capture new galaxies over time, which may be transformed as they traverse the cluster environment. Visible in this image is a rather blue, spiral galaxy that is forming stars (in comparison to the 'red and dead' ellipticals and lenticular galaxies in the cluster). Understanding galaxy evolution in clusters is an important area of current research.

GALAXY

But we do so from a very limited vantage point. On a cosmic scale, we humans effectively inhabit a two-dimensional membrane: the surface of the Earth (and the thin, few-hundred-kilometre-thick layer of space surrounding it, as well as a few well-placed satellites in extraterrestrial orbits). In any case, we are basically trying to audit the contents of the entire universe from a single point within it. This makes the job far more difficult than if we could move our vantage point around arbitrarily. Alas, the laws of physics forbid this luxury.

The first challenge we face as cosmic cartographers is that we can only measure the positions of galaxies within a spherical coordinate system, defined by the positions of galaxies on the sky (the inside surface of a sphere) and a redshift (or, if we're lucky, a 'proper' distance measurement, like the parallax or Cepheid method, but generally these only work locally), which is a radial distance outwards. Mapping the local volume isn't too hard because most of the galaxies are bright and easy to measure. Nevertheless, it's still easy to miss nearby galaxies that are small, or have very low luminosities, and occasionally we will find a new member of the Local Group – a nearby galaxy that has eluded detection until recently.

As we probe further out into the universe, the apparent sizes of objects get smaller and fainter, and this makes them harder to observe. A survey that is limited by 'depth' (that is, a short exposure time or low sensitivity) starts to miss out galaxies that are too faint to be detected by the camera, or whatever instrument we're using. We call this 'incompleteness' in surveys, and we must recognize and understand it if we want to avoid jumping to the wrong conclusions in our analyses. For example, imagine standing on that hill in chapter One, looking out to the distant horizon and seeing the glints of other cities in the distance. The distant cities are quite easy to spot, but you might not be able to see the distant towns and villages that don't boast huge skyscrapers. You might conclude that there are no other villages and towns – all that's out there are large metropolises. But this would likely be wrong: you are unable to detect the distant villages and towns, but that doesn't mean they're not there. Instead, it would be more prudent to assume that, since your own city has several villages clinging to the outskirts, those distant cities – which look about the same size as yours – have about the same number. We play similar games with observations of the distant universe; we have to make assumptions about the stuff we cannot yet see, and make predictions for when better instruments come along that will be able to confirm or refute our hypotheses.

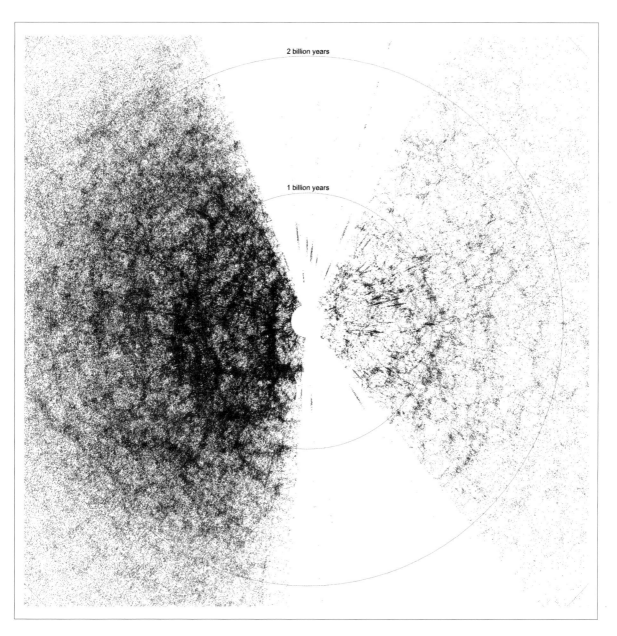

2 billion years

1 billion years

A map of the local distribution of galaxies identified in spectroscopic redshift surveys to date (the key one being the Sloan Digital Sky Survey), plotted with Earth at the centre and cosmic distance increasing radially. The two rings correspond to light travel times of 1 and 2 billion years: when we look at distant galaxies, we are seeing them as they were in the past, providing a means to study the properties of galaxies earlier in the history of the universe. The two wedge-shaped regions with few galaxies are the Zones of Avoidance, where the disc of the Milky Way is too thick to allow light from extragalactic sources to penetrate. Note how the galaxies are distributed into a foam-like filamentary structure – the 'cosmic web' of matter.

Another problem, as we discussed in the first chapter, is that we never get a complete 'all sky' view of extragalactic space; the disc of the Milky Way is so thick with stuff that pretty much no light from distant galaxies can make it through. The band of the Milky Way across the sky is called the 'Zone of Avoidance'. When we see maps of the galaxy distribution, they are usually wedge-shaped, reflecting the fact that we can only see distant sources clearly in the bands above and below the plane of the galaxy, where the density of stars, dust and gas is low. While inconvenient, this is by no means a disaster. For one, being actually embedded within the disc of a galaxy allows us to study it in great detail on spatial scales that are – for obvious reasons – impossible to achieve for external galaxies. This keeps 'galactic astronomers' busy with observations of the inner workings of the Milky Way, with most of the active research focused on the action-packed galactic plane. Second, there is a cosmological tenet called the 'principle of isotropy', which states that on large scales the universe pretty much looks the same in all directions. That is to say, as long as we observe a decent enough chunk of the universe above and below the disc, we can be pretty sure that if we could see through the galaxy, then the galaxies in that direction would (statistically) be pretty much the same. To be succinct, we're not missing anything. Put another way, if we took our 1-cubic-metre box and placed it down in some completely random part of the universe, we'd find that, although the exact layout of galaxies might be different, there would be a similar number of galaxies, groups and clusters, and that their statistical properties would be the same.

The progress we've made in mapping the universe to ever larger scales necessarily started slowly. In prehistoric times humans first noticed the stars in an intellectual way, starting our astronomical journey, but limiting human knowledge of the universe to the confines of our own galaxy. Not a huge amount of progress was made for a very long time because our technology didn't improve – there's only so much the human eye can do. But in the last 400 years, since the invention of the telescope by Dutch opticians in the early seventeenth century, we've been able to explore far further, expanding our horizon. There is no question that this has been completely driven by the technological advances and innovations in telescope and detector technology. This progress continues today at a faster rate than ever before. As I write this, there are plans to build 'extremely' large telescopes, with primary mirrors three or four times larger than the biggest (visible light) telescopes today. We can even put telescopes in space and control them remotely from Earth; just imagine what the early telescope pioneers would have thought of that! Similarly, new instruments are always being developed that are more sensitive, efficient, ingenious or technologically advanced than previous designs. This keeps our field fresh and exciting, since there is always the promise of a completely new discovery waiting to be made: there is always the potential to see further, or in better detail, provided we have the technology.

If we take an image of part of the sky and detect a galaxy, provided we can measure the spectrum of that galaxy, we can normally measure its redshift, or at least have a good guess, and therefore place that galaxy in some three-dimensional model of the universe. The position on the sky gives us two of the coordinates, and the redshift gives the third. This gets more and more challenging for very distant – and so very faint – galaxies, because measuring an accurate redshift, and indeed detecting the galaxy in the first place, requires us to collect enough light so that the astronomical signal is sufficiently larger than the random noise due to electronics, the ambient thermal background and so on. Random noise, not associated with the signal we are trying to detect, is present in all electronic detectors. We are also limited by resolution. If you are in a field full of cows, the ones closest to you look bigger than the ones further away. If you take a digital photograph of those cows, then the more distant ones occupy fewer pixels on the image than the ones in the foreground. We can see more details in the cows closest to us, but the ones on the horizon might only just be distinguishable as cows at all. The same goes for galaxies. Nearby galaxies are easy to spot since they appear large on the sky, and we can make out internal details like spiral arms, bars, bulges and even individual star clusters and star-forming regions. More distant galaxies appear smaller, and because our instruments have a limited resolution (that is, the smallest angular scale that can be discerned, which is fixed by the size of the telescope), in most cases we can't make out *any* details at all – the galaxy is just a collection of a few bright pixels in our image. In fact, if we really push it, we get close to the limit where we have to be careful about mistaking a collection of bright pixels – which might be a really distant galaxy – for a spike of random noise. Usually we require a follow-up observation to confirm or refute the reality of such systems. If the noise is random, then it is highly unlikely that we'd get another noise spike in exactly the same position in the image, so re-detection of some faint, putative galaxy in an independent image is more compelling evidence than a single exposure alone.

We generally only trust an astronomical detection, be it a simple image of a galaxy or some feature in its spectrum, when the signal we see is at least five times the typical size of the random variations due to the noise in the measurement (for example, electronic noise in a CCD image). Our ability to 'beat down' the noise level (by making more and more sensitive cameras and detectors), collect as much light as possible (so we can gather that tiny flux of photons from a distant object) and cover large areas of sky (so we can survey as much of it as possible in an efficient way) are the three magic ingredients in our quest to map out the universe. All three are driven by technology: we want the most sensitive detectors assembled into big cameras and mounted onto large telescopes.

A significant part of the observational effort in galaxy studies for the past half century has revolved around sky surveys, and in fact now they are more important then ever; it is often said that we are in a golden age of galaxy surveys as it becomes easier to perform extremely large-area, sensitive sky surveys with a range of instrumentation. Surveys are useful not only for mapping out the locations of galaxies in the universe, which – as we have seen – are not random, but also for amassing large samples of galaxies of different properties and, most importantly, thanks to the fact that light takes so long to traverse cosmological distances, seen at different times in the universe's history. Look back far (that is, faint) enough and you will see the light emitted by the first galaxies fairly shortly after the Big Bang. This is how we can investigate how the main galaxy properties like stellar mass, shape, chemical composition and so on evolve over time.

Perhaps the most successful galaxy survey to date has been the Sloan Digital Sky Survey, or SDSS, which began operations in 2000. With a relatively small, 2.5-metre telescope, located at Apache Point Observatory in New Mexico, the SDSS has spent the past decade imaging about a quarter of the whole sky. It has produced arguably the best map of the local universe that we have. At its heart, the SDSS has a large, 120-megapixel CCD camera. This can take an image of one-and-a-half square degrees of the sky, which is pretty big; remember, the size of the full Moon on the sky is half a degree across. This large field of view allows the SDSS to quickly build up the survey area; in fact, the imaging technique is slightly different to many telescopes. Instead of aiming at a particular position and taking an exposure, the SDSS employs 'drift scanning', which uses the fact that, as the Earth rotates, the stars appear to drift across the sky. If you place a telescope on the ground, pointed

This image is constructed from the positions of all the galaxies detected by the Sloan Digital Sky Survey, revealing the overall projected density of galaxies in a large region of sky called the Northern Galactic Cap. You can see that the galaxies are not randomly distributed: there are patches of high density (the clusters), and clear filamentary structures forming a network throughout the galaxy distribution. This is the 'large scale structure' of the universe, with galaxies forming and evolving within an unseen skeleton of dark matter structures that have evolved over time through the force of gravity.

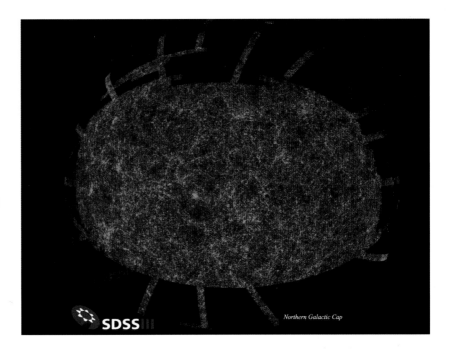

Northern Galactic Cap

up, then over the course of a night you will record a strip of sky as the Earth turns. Thus the SDSS images the sky in a series of strips. One advantage of drift scanning for large-survey work is the accuracy gained in 'astrometric' calibration (how well we can translate pixel positions on the resultant image to the actual positions of sources in the sky). The SDSS is a relatively 'shallow' survey in that, to image such a large area of sky, it cannot achieve the long exposures that can probe down to very faint galaxy fluxes, compared with, say, the Hubble Ultra Deep Field, which was a dedicated long-exposure observation of a tiny patch of sky by the HST, which revealed extremely distant galaxies. Most of the galaxies detected by the SDSS are fairly local by comparison. On the other hand, the sheer scale of the survey means that the size of the cosmic volume probed by the SDSS is huge, and this is what is so useful.

A multicoloured view

The SDSS takes images through five different coloured filters, called u, g, r, i and z, spanning the full visible light spectrum from blue to red. These are examples of the broad-band filters we talked about earlier in this chapter – they are designed to only allow light through with a certain range of wavelengths. It's important to have these different filters because, as we saw earlier, the galaxies can have different shaped spectra. These spectra, remember, correspond with the amount of energy being emitted at different wavelengths. Some galaxies are emitting more blue light, for instance, and this becomes apparent in an image taken with the u or g band filters, since these are 'sampling' the blue part of the galaxy spectrum – the galaxy will appear brighter in these filters compared with, say, the z band.

In short, different galaxies can look different when viewed through filters of different wavelengths. A galaxy that appears brighter in the r band compared to the g band, say, is said to be 'red'. Conversely, a galaxy that is bright in the g band and fainter in r is said to be 'blue'. This use of galaxy 'colour' is a beautifully simple classification system; as a rule of thumb, blue galaxies are actively forming stars, because – as we discussed earlier – the blue light is dominated by emission from recently produced young, massive stars that are bright at ultraviolet and blue wavelengths. Once star formation has stopped, the blue stars die off, and older, mature stars dominate the spectrum, leading to a 'red' galaxy. Red galaxies are often called 'passive', or 'red and dead', but remember we have to be careful, because dusty, actively star-forming galaxies can also appear red, mimicking a passive galaxy. Similarly, the more distant galaxies also appear redder, since their light is redshifted to longer wavelengths; therefore we have to make corrections when comparing the broadband colours of galaxies at different redshifts, even if they are of similar type (say, two spiral galaxies).

Comparing the amount of light coming from each of the u, g, r, i and z bands can also be used to estimate the galaxy's redshift, because what we are effectively doing is measuring a crude spectrum – not as detailed as could be achieved with spectroscopy, but enough to get the overall shape. In the case of these five SDSS bands, we get a measurement of the average amount of energy emitted by the galaxy at wavelengths defined by each of those filter band passes: we have the general shape of the 'continuum' of the spectrum, but lack any fine detail (like emission lines). By comparing the relative fluxes in each of these bands with that expected for a template, or model, spectrum, we can estimate things like the type of stellar population (old or young on average?), the total mass of the stars and, importantly, the redshift of the galaxy. These 'photometric' redshifts, although far inferior to those measured with spectroscopy with very high precision, are extremely valuable because they are so much 'cheaper' in terms of the amount of observing time you have to invest compared with obtaining a spectrum. Why is this? When measuring a spectrum of a galaxy we are in a sense diluting, or smearing out, the amount of energy that hits a given pixel in our detector because we have dispersed the light, separating it out into its frequency components so we can examine it in detail. This comes at the substantial cost of increased exposure time compared to a simple two-dimensional image of that same galaxy using a set of broadband filters, which allows lots of photons through, building up a signal in the detector very quickly.

Despite the cost of spectroscopy compared with imaging, since it is so useful, with a big survey like the SDSS, there *are* methods to make the collection of spectra very efficient. In addition to the imaging component, which has catalogued over half a billion objects so far, there is also a spectroscopic component to the SDSS, which uses a 'multi-object spectrograph': a spectrograph that can measure the spectra of many galaxies simultaneously. The SDSS spectrograph is a 'fibre-fed' device, meaning that it obtains spectra by placing an optical fibre in the path of light coming from a galaxy. In practice, the SDSS images a patch of sky first to identify targets to measure spectra for, since you need to know in advance where to put the fibre. Once targets are selected, an aluminium sheet, or 'plate', can be placed in the focal plane, with holes drilled in it at the positions of the desired targets. The ends of the fibres are placed in the holes, and these intercept the photons from each target, redirecting the light down to a dispersion element which splits the light from each target into its spectrum. The SDSS can measure spectra for over 600 targets simultaneously, and to date it has measured the spectra and redshifts of millions of astronomical sources. Incidentally, the data produced by the SDSS is public – anyone can download the images and catalogues produced by the survey and explore the universe – and there are regular 'data releases' as the survey progresses.

A type of galaxy that the SDSS has spent a lot of time obtaining spectra for are quasars, or 'quasi-stellar objects' (a.k.a. QSOs). Quasars are a seriously active class of galaxy; they are among the most luminous systems in the universe. Because of their high luminosities, quasars are visible over huge cosmic distances, shining out like beacons. Although the SDSS's galaxy catalogue is limited to a fairly local volume, its quasar catalogue extends to the extremely distant universe.

Essentially quasars are just galaxies, but their difference from a typical galaxy, like the Milky Way, is in the amount of energy emitted by their core, or nuclear region. The light emitted by a quasar's nucleus is so immense that it outshines the rest of the galaxy. In fact, the light is so concentrated and intense that quasars appear as single points of unresolved light (we often can't distinguish spatially extended features in the galaxy), like a star, hence their name 'quasi-stellar'. What drives this power? Quasars contain, at their centres, a growing supermassive black hole. The supermassive black hole is so called because it is much more massive than the black holes produced by single stars, which can form at the end of certain (massive) stars' lives. The *supermassive* black holes can be millions of times more massive than the Sun. Although they probably started out much smaller (perhaps merging with central black holes in other galaxies), they grow over time within the galaxy by swallowing material, mostly interstellar gas and dust. It is this accretion of material that is the power source of quasars. As the black hole accretes gas and dust, it forms a dense, compact disc. Because of the enormous gravitational and dynamical forces involved, this disc gets so hot that it glows brightly with x-ray, ultraviolet and visible light.

Another name for this region within the quasar is the 'active galactic nucleus', or AGN. Sometimes astronomers refer to galaxies as AGNs even if they are not classified as full-blown quasars, because the nuclear emission is dominant over the galaxy as a whole. Actually, most galaxies contain a supermassive black hole at their heart, and our own Milky Way is no exception. Observations with the Very Large Telescope have actually tracked, over a period of a few years, the orbits of stars around the Milky Way's central black hole (which, on the sky, can be located towards the constellation of Sagittarius). Although the black hole itself, and its environs, are not visible, the shapes of these stars' orbits imply the presence of a massive, dark object.

Our Milky Way's black hole is not really 'active' – it's not swallowing matter at a high rate and releasing huge amounts of energy. Occasionally something *will* fall in – as I write this, there have been observations of a gas cloud on its way to being swallowed, at which point there should be a brief release of energy as the gas ploughs in. Telescopes are prepared for this unique opportunity to study the accretion of gas onto a supermassive black hole in a local setting. In AGNs and quasars, this accretion is happening all the time, and understanding the physics of this process and how it fits into the global scheme of galaxy evolution is a key area of current research. Although quasars are bright in the visible light bands, astronomers also search for actively growing black holes in galaxies using x-ray telescopes. X-ray observations can only be conducted from space, since these high-energy photons

Quasar MC2 1635+119, imaged by the Hubble Space Telescope in a single band of light. The centre of this galaxy shines like a star: most of the light is coming from the very central, nuclear region where a supermassive black hole (present in all massive galaxies) is actively accreting matter, throwing off huge amounts of energy in the process. This activity could have been triggered by a galaxy–galaxy merger, which can cause gas to be forced to the central region of the merged galaxy, compressed to high concentration. Here it can readily feed the growing black hole. The faint, fuzzy emission around the central source shows the disturbed nature of the stars in the 'host' galaxy, which could support this picture. Quasars are so bright that they can be seen over great cosmic distances, and so are therefore excellent probes of the early universe. They also have an important role in massive galaxy evolution, since the intense nuclear activity can affect the star formation history and therefore the future destiny of galaxies such as this.

Markarian 509, a galaxy with an active nucleus, visible as the bright unresolved point of light at the centre of this image. The nuclear activity is due to a supermassive black hole (several hundred million times the mass of the Sun) that is actively accreting matter. As material falls towards the black hole, it forms a hot accretion disc that glows brightly with x-ray, ultraviolet and optical radiation, sometimes outshining the rest of the galaxy. Every massive galaxy harbours a supermassive black hole, and the mass of this central 'heart' is observed to be correlated with the mass in stars in the surrounding stellar bulge. It is thought that the black hole and bulge growth are linked via regulatory feedback mechanisms, and understanding this astrophysics is a key area of research.

This is the very heart of our Milky Way, imaged with a special technique called 'adaptive optics' that can correct for the blurring effect of Earth's atmosphere, which usually limits the spatial resolution of images that can be taken from the ground (one of the reasons why the Hubble Space Telescope can produce such exquisite images is that it does not have to contend with the Earth's atmosphere). In this case, the sharp image made possible by adaptive optics allowed astronomers to pinpoint and, over a sixteen-year period, monitor the positions of stars actually orbiting an unseen massive, compact object – a supermassive black hole – that lurks at the heart of the galaxy. All massive galaxies contain one. In quasars and active galactic nuclei the central black hole is actively accreting matter, causing the region to glow brightly, but in our case (as is the case in the majority of galaxies), the Milky Way's central black hole is relatively 'quiescent'. However, it still has a gravitational influence on the stars around it, and by measuring the orbits of several stars in this image, astronomers could determine the mass of the black hole, which is of the order of a million times the mass of the Sun. The centre of our galaxy is about 8,000 parsecs away; if we scaled down the distance between the Earth and the Sun to just 1 millimetre, the distance from the Earth to the centre of the galaxy would be 1,600 *kilometres*.

cannot get through our atmosphere. Two important x-ray observatories of recent years have been the XMM-Newton (XMM stands for 'X-ray Multi-mirror Mission', and the mission is also named after Newton) and Chandra (named in honour of Indian-American astrophysicist Subrahmanyan Chandrasekhar, who was an important contributor to twentieth-century astronomy) satellites. These telescopes provide some of the highest-energy views of the universe, and therefore insights into the most extreme astrophysics that occurs in galaxies.

The intensely active nuclei of quasars and AGNs are copious emitters of x-rays, where the level of x-ray luminosity is directly linked to the accretion of the central black hole. So surveys with telescopes like XMM-Newton and Chandra can find and characterize these systems (although often only a handful of x-ray photons are detected from a given system). Just as with optical light, however, AGN are often enshrouded by thick dust, which can obscure the x-ray emission. Luckily, just as we can identify dust-enshrouded star-forming galaxies in the infrared, we can do the same to identify dust-enshrouded AGN, where the screen of dust has been heated by the energy emitted by the hot accretion disc, thus giving off a distinctive infrared glow. The SDSS has obtained spectra for hundreds of thousands of quasars so far, and these are some of the main 'tracers' of the galaxy distribution of the furthest reaches of the universe.

The problem of distance

There is a slight catch with mapping the universe with redshifts, since observed redshifts are not quite the same thing as true distances. Hubble's law tells us that there is a correlation between redshift and distance: things with higher redshifts are further away. This means that – if we don't have a direct handle on the actual distance – a

redshift provides an easily measurable proxy. But galaxies are not just going with the expansive 'Hubble flow' of the universe; they are also in motion because of the relentless gravitational attraction of other galaxies and matter in the universe. So, in addition to their relative motion away from us due to cosmological expansion, there is an extra component to their motion caused by local gravitational effects. This is called 'peculiar velocity'.

The magnitude of the peculiar velocity of a given galaxy depends on the distribution of matter around it. For example, galaxies in big clusters have very large peculiar velocities, up to 1,000 kilometres per second or so, because they reside within, or close to, a very large mass concentration which forms a gravitational 'potential' that can accelerate them to high velocity relative to other galaxies in the cluster. A galaxy at the edge of a cluster is like a bowling ball at the top of a steep hill – release it and it will accelerate to the lowest point of the potential 'well'. If it has enough energy it will start shooting up the other side of the hill, and so forth. This is analogous to a galaxy on a 'radial' orbit around the cluster core. Galaxies in clusters are doing this all the time – zooming about like a swarm of bees because they are orbiting a common mass. In fact, collectively, the distribution of relative velocities of galaxies in a cluster can be used to estimate the total mass (including all of the dark mass) of the cluster, since the range in velocities is linked to the mass enclosed within the system. In practice, instead of measuring all the cluster galaxies' velocities relative to the Milky Way, we compare their velocities to the average redshift of all the galaxies in the cluster. When we plot the distribution of 'Delta V' for all the galaxies in a cluster, we find a classic bell-shaped or Gaussian curve. The characteristic width of this distribution is called the 'velocity dispersion'. If we know the size of the cluster, which is of the order of one to a few megaparsecs in diameter, we can estimate the cluster's total mass.

The large peculiar velocities of galaxies in clusters is a nice demonstration of how, when we try to place galaxies in a three-dimensional model of the universe, we do not quite get an accurate view. Think back to our box, with the Milky Way in the middle. Looking out from the Milky Way, we can measure the positions of galaxies on the sky very easily by just taking an image. The problem comes when we need the third dimension, because we can only measure the redshift in a radial direction. So, in clusters like the Virgo cluster, the redshifts of all the individual galaxies are affected by a significant extra component of velocity in addition to the general recessional

The Chandra x-ray Observatory being deployed in orbit by the Space Shuttle. Chandra has been one of the key satellite observatories of recent years, providing a window onto the most energetic processes in the universe, particularly the x-ray emission associated with growing black holes in distant galaxies.

velocity caused by the expansion of the universe, due to acceleration in the cluster's gravitational potential. What this means is that we do not know exactly where in the cluster those galaxies are: we are looking at the galaxies in 'velocity space', not true space. This is evident if we just plot their positions along the line of sight as given by their individual measured redshifts; we get what looks to be an elongated thin clump which is a result of their large relative velocities compared with other galaxies at the same distance from us, but away from the cluster and not affected so much by its gravitational influence. In reality, in real three-dimensional space, the galaxies in clusters are distributed in (usually) a symmetrical spherical halo. We can see this easily from the two-dimensional layout of the galaxies on the sky, but the spatial information is lost in the third, radial dimension. In surveys, these elongated clumps became known as 'Fingers of God'. This effect is inconvenient, but not a disaster. Astronomers have come up with clever ways to compensate for these so-called 'redshift space distortions' when making cosmological measurements from redshift surveys.

Measuring the true distance to objects is the hardest problem in astronomy. It gets even harder the further you try to look, because methods that work for nearby objects become impossible. Parallax measurements only work for a relatively small bubble of space within our own galaxy, stretching no more than a few tens of parsecs from Earth. Using Cepheids as distance indicators works fine as long as you can pinpoint the individual stars, but again, as we look to more distant galaxies this becomes harder and harder as all the starlight from a galaxy gets blended together and we can no longer 'resolve' those stars. This limits Cepheid variable observations to galaxies in our local volume; most things within our 1-metre box representation, but not much beyond. But there is one very special circumstance that we can take advantage of to extend our reach: we *can* use individual stars as standard candles even in the very distant universe when they explode as supernovae.

Supernovae are the events that mark the violent ends of certain massive stars (not all stars are capable of 'going' supernova; they have to be above a certain mass threshold). There are two main types of supernova, but the one of interest here is called Type 1a. Type 1a supernovae occur where one of the stars in a binary system (two stars in orbit around each other) comes to the end of its life and collapses into a compact object called a white dwarf. This collapse happens when nuclear reactions in the core can no longer hold the star up against the force of gravity, which is always trying to crush it into oblivion. All that is holding the white dwarf up against total collapse is a kind of pressure that occurs due to quantum effects between electrons within the ultradense matter that is the remains of the star. This is the Pauli exclusion principle at work, which states that two fermions (an electron is a fermion) cannot share the same quantum state. However, new material can be accreted onto the white dwarf from the nearby companion star, increasing the pressure in the core of the remnant to a critical limit. After enough new mass has been accreted onto the white dwarf, pressures and temperatures increase until a threshold is passed, whereby carbon and oxygen nuclei in the white dwarf suddenly fuse. This causes an explosive reaction, shattering the star. Enough energy can be released in these explosions to briefly outshine the rest of the galaxy, and they are visible over vast, cosmological distances.

To detect a supernova, all you need to do is take an image of a patch of sky, wait a bit, say a week, then take another image of that same patch. Repeat this as many times as you want. The larger you can make your image the better, because it will contain more galaxies. Normally two consecutive images will look identical, because the galaxies have not moved or otherwise changed in appearance on the sky. The only difference should be the conditions of the observations (for example, maybe one night was a bit cloudier than another, or sunlight glinting off a satellite or the light of a plane may have left a trail in the image), the effects of which are easily detectable and removed. But once in a while something will be different – there will be a bright spot within or near to a galaxy where there was not one before. This is a classic sign of a supernova. As I write this, a supernova has recently gone off in the galaxy m95, and astronomers – professionals and amateurs alike – are frantically turning their telescopes to monitor it. When a supernova goes off in a well-known galaxy like m95, it's very obvious, but the same is not true of most galaxies.

Once a supernova goes off it rapidly brightens to a maximum peak, then begins to fade over a period of days and weeks. This is called the supernova's light-curve. The fading light in a Type 1a supernova is caused first by the radioactive decay of the element nickel, which has a half-life of about a week (which means that, in one week, about half of the nickel has decayed into other isotopes), and then later by the decay of cobalt, which has a longer half-life of about eleven weeks. So, the fading light of a supernova is visible for a fairly long period of time, and thus can be tracked, but it is essential to capture the supernova as close to the peak as possible, then observe at regular intervals to properly measure the process of the fading and to get a good measurement of the shape of the light-curve. In addition, supernovae are pretty rare events in normal galaxies – at least, on human timescales they are – occuring on average about one per century per galaxy. The best chance of catching one going off is to observe large numbers of galaxies: if you monitor 100 galaxies, say, you might detect one supernova per year. Observe a million galaxies and you could detect about 30 per day, provided you are careful with your monitoring and detection algorithms (a million galaxies are too many to inspect by eye – you need a computer to do it for you). Again, this is where very large surveys come in handy.

Here's the important bit: it is thought that all Type 1a supernovae have the same intrinsic luminosity at their peak. As we have seen, if you know the true intrinsic luminosity of an object, and compare this to the brightness we actually detect, you can work out how far away it is. In other words, supernovae (Type 1as, that is) are standard candles, like Cepheids. This is fantastically useful, because it allows us to obtain distance measurements for galaxies far beyond the local volume and thus calibrate Hubble's law over cosmological distances.

Over the past decade, two teams of astronomers led by Saul Perlmutter and Brian Schmidt made a consolidated effort to detect and measure supernovae in large samples of distant galaxies. But when the data were actually put on the Hubble plot of distance versus redshift, a surprising discovery was made: distant supernovae appeared fainter than would be expected if one were to do a simple linear extrapolation of Hubble's law as measured for the local volume. What could this mean? One explanation for the dimness of the distant supernovae is that they are actually more distant than is predicted by a naive extrapolation of Hubble's Law. The supernovae results suggested that the rate of the expansion is actually accelerating, which means that the supernovae appear fainter at a given redshift. The name given to the origin of this acceleration is 'dark energy', the exact nature of which is unclear. We're not going to talk too much about dark energy here, because it does not impact too greatly (at the moment) on the evolution of individual galaxies. Needless to say, the research and discovery of the accelerating universe was so important that Perlmutter and Schmidt, along with Adam Riess, who was one of the major contributors to the discovery, were awarded the 2011 Nobel Prize in physics.

From a cosmologist's standpoint, supernovae are useful events to be used as tools in understanding the geometry and expansion history of the universe.

Supernovae also have an important role in galaxy evolution. Without them, we probably wouldn't be here. The key is their explosive power. Stars are nuclear furnaces in which most of the elements that were not made in nucleosynthesis shortly after the Big Bang are formed. Stellar nucleosynthesis happens through the process of nuclear fusion in stars' cores, where lighter elements are fused together into heavier ones. The nuclear reaction releases energy, which we see as starlight. We've been trying to make nuclear fusion work on Earth as a practical power source for years, mimicking the physics of stars, but making fusion work on an industrial scale is a tremendous technological challenge. It will happen, but perhaps not for several decades, and for now we have to make do with fusion's dirtier cousin, fission.

When a supernova goes off, the explosion expands rapidly outwards like an inflating balloon, smashing into whatever is in the vicinity, sweeping up the heavy elements and dispersing them into the surrounding space. Over time, through the continual detonation of other supernovae (the rate of supernovae in a given galaxy is related to the rate at which that galaxy is forming new stars), the interstellar medium is polluted with new elements – the 'metals' – that were formed in the stars. The blast waves of supernovae, along with the winds that are blown from the surfaces of stars and the rotation or other internal motions of the galaxy itself, cause this pollution to be mixed around. We call this process enrichment.

A cloud of hydrogen that has been enriched with metals can go on to collapse again to form new stars. Hydrogen is extremely abundant, and is not used up all 'in one go' – so star formation can be sustained in galaxies for fairly long periods of time. The stars that form within such a cloud will be more 'metal rich' than the previous generation. In addition, when those new stars are born, they also form and are surrounded by dusty discs. This is the genesis of new solar systems; within these dusty discs new planets can form. Our own solar system formed in the same way. Planets like the Earth are made primarily of iron and silicon and, as we know, the Earth contains many other elements – the most important for us being things like carbon and oxygen, which make life as we know it possible.

When our own Sun dies around 5 billion years from now, when it burns up its hydrogen fuel, it will not go supernova. It is not massive enough. But it will evolve into a 'red giant' star that will expand outwards, engulfing and incinerating the inner planets, and most likely will destroy, or at least seriously affect, the outer gas giants. Eventually, through its death throes, the shedding

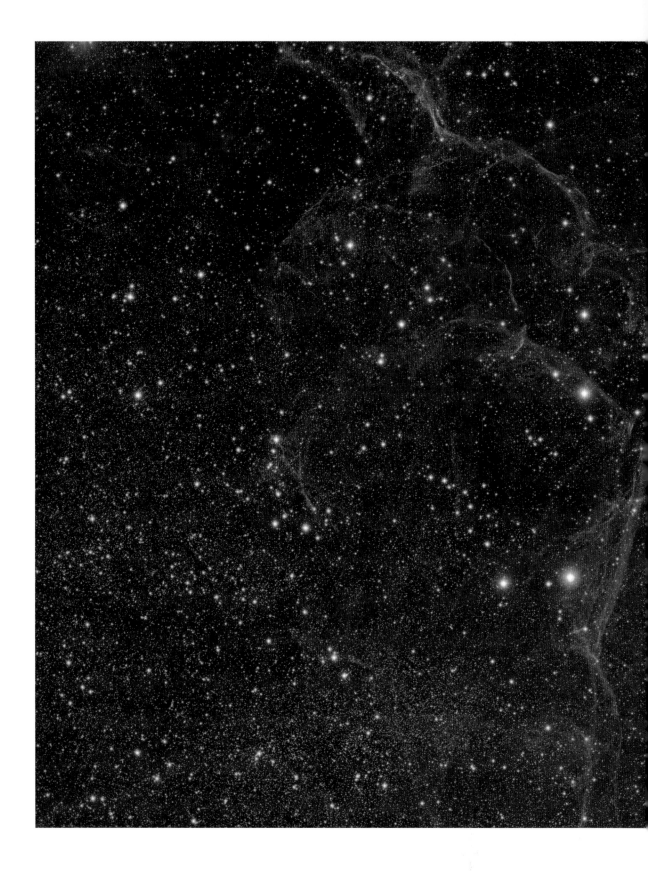

The delicate pink and blue nebulosity in this image is the Vela supernova remnant, the shattered remains of an exploded star in our galaxy (the explosion happened about 10,000 years ago in this case). Supernovae are the violent deaths of massive stars, and have a role in distributing the material formed during stellar evolution throughout the interstellar medium, as well as 'dumping' energy into their immediate vicinity as the explosion rams into surrounding gas and dust. This can clear out some of the gas around star-forming regions and therefore help regulate new star formation by controlling how much of the dense gas can gravitationally collapse. The combined effect of many supernovae exploding in a galaxy can be a galactic 'wind' that can drive the gas and dust out of the galaxy. Supernovae are so bright that they can be observed across cosmological distances.

A glowing shell of gas that is the remnant of a supernova, SN0509-67.5: another demonstration of how elements generated within stars can be dispersed into interstellar space.

This near-circular shell is the supernova remnant SN 1006, imaged in radio (red), visible (yellow) and x-ray (blue) light. The image shows the expanding shell of hot gases that were blown outwards by the explosion of a star in our galaxy (x-rays show the emission from the hottest gas). The star exploded about ten centuries ago, and now the products of stellar evolution – the heavy elements, as well as other elements forged in the explosive event itself – are being dispersed back into the interstellar medium. Supernovae thus have a role in enriching the interstellar medium of galaxies, to be incorporated into new generations of stars (where heavy elements can form things like planets and people). Their explosive power also deposits energy into the interstellar medium, and this feedback can drive powerful 'galactic winds' that can transport material away from the sites of star formation and, in extreme cases, the disc of the galaxy itself (see M82).

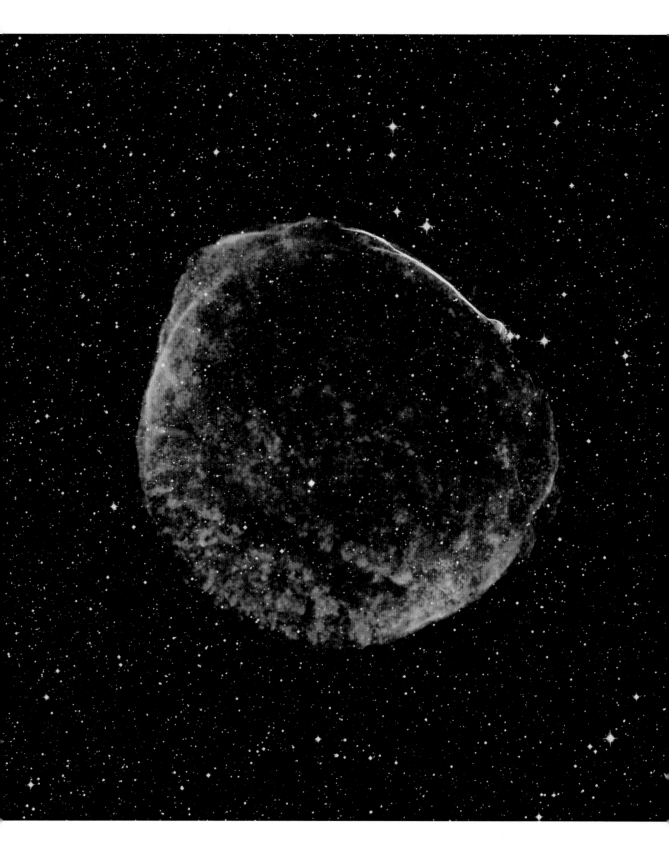

of its outer layers, the Sun will enrich our solar environment with a new generation of heavy elements, 10 billion years in the making, increasing the metallicity of the Milky Way by a little bit. One day, far in the future, some of this new material might find its way into a new solar system, and maybe a brand new ecosystem. Humans can escape the fate of the solar system by mastering interstellar travel, or at least the capability to voyage to a safe location between the stars. We have 5 billion years to figure it out.

The continual collapse of gas clouds into new stars, and the recycling and gradual enrichment of the interstellar medium through stellar evolution, is a key process in galaxy evolution. My work involves studying galaxies at large redshift far beyond the local universe; way beyond the edges of our 1-metre box. One of the key concepts I want to convey to you is that although we have talked about the connection between redshift and distance, I do not think of the galaxies that I study as being very far away in space (although they are). I think of them as being distant in time, not contemporary with the Milky Way we live in. The reason for this is that the distances between are so vast that the light we see emitted by

'distant' galaxies was actually physically emitted in the ancient past – billions of years ago. So what we're actually measuring is what galaxies were like when the universe was younger. Those distant galaxies are now (at this instant) different, but we can't see them as they are right now because this light has not yet reached us.

This is no inconvenience, though, because by looking into the past we can understand how the cosmos and its contents have changed and evolved. By looking to ever more distant galaxies we're looking further and further back into the past. This is the essence of my field – the field of galaxy evolution.

Let's very briefly recap before continuing on our journey. We are trapped on a planet, orbiting a star that itself is orbiting (along with billions of other stars and solar systems) a disc-like stellar system we call the galaxy, our Milky Way, filled with stars and gas and dust, and – at its heart – a huge black hole. We know that there are other galaxies outside our own, some like the Milky Way, some not, separated by vast distances and organized into a large-scale structure of clusters and groups and filaments. As the stars burn in these galaxies, new elements are formed; metals that can be dispersed throughout the galaxy as the stars die. As the distances between galaxies get very large – what, in the trade, we would call 'cosmological' – then the time it takes light to traverse the great distance between far-flung galaxies and our detectors becomes significant, so we're seeing those galaxies as they were at an earlier time. When we look at very distant galaxies, we're getting a snapshot of the universe as it was in the past, and this is the basis for our studies of galaxy evolution. So far, so good.

The end of a star: the Cat's Eye nebula in the Milky Way. This extremely complex structure is the remnant of a star that has shed its layers during the final stages of stellar evolution, giving rise to what is known as a planetary nebula. A compact white dwarf, what remains of the core of the star, is forming at the centre. The end-phase of stellar evolution – the deaths of stars – is an essential part of the overall evolution of galaxies, since it allows heavy elements produced *inside* stars during their lifetimes to be dispersed into the interstellar medium. These heavy elements, or metals, 'enrich' the interstellar medium and become intermingled with new generations of stars. No better evidence of this is our own solar system: the existence of planets and people relies on the fact that the cloud of gas the Sun formed from was polluted with the ashes of dead stars. The Cat's Eye is a glimpse of the fate of our Sun in around 5 billion years.

The Orion nebula is a region in our galaxy where new stars are being formed, here imaged in near-infrared light. Orion glows with the light of ionized gas and scattered starlight as the new stars illuminate their gaseous and dusty birthing grounds. Imaging this region in near-infrared light allows astronomers to peer through much of the obscuring dust that cocoons the emerging stars at the heart of the nebula.

THREE

Seeing More

Constellations and asterisms are not (necessarily) physical associations of stars, but give that illusion because we cannot make out their three-dimensional distribution with the naked eye. What we see is just the projection of bright stars at different distances that happen to line up to make interesting and recognizable shapes. The ability of the human brain to identify patterns in the stars was our species' first foray into the science of astronomy, because it allowed us to create maps of the stars, dividing up the sky into interesting regions. Without these handy patterns it would be much harder to re-identify or give the location of a given star, planet or comet.

Next time you can see the constellation Orion, take some time to really examine it in detail. It's my favourite constellation, not only because of the unmistakable belt, that line of bright stars that seems to instantly trigger the primitive pattern-recognition functions of our brains, but also because within that constellation is the famous nebula – the Orion nebula – a huge gas complex that is forming *new* stars.

The Orion nebula is big and bright enough that a normal pair of binoculars will easily reveal it, and at the darkest sites you can even make it out with the naked eye. In my opinion, the finest telescopic views of the nebula are among the most spectacular images of the natural world we have. The nebula blazes with the light from ionized gas – hydrogen mainly, which has a reddish glow – as well as the blue light emitted by nascent stars that gets reflected and scattered off the interstellar gas and dust. There is also emission from other elements mingled in with the nebula. Among the Turneresque scene there is, of course, the famous Horsehead nebula: a filigree of dense gas that actually absorbs the light being emitted behind it, causing it to stand out in equine silhouette.

The ability to actually see this star-forming complex from my own back garden when I was younger, using my little refracting telescope, was one of the factors that inspired me to become a professional astronomer. At the time I was unaware of the detailed astrophysics governing its glow, but the very fact that it is something you can see with your own eyes – not in some book – captured my imagination. It was the first time I *really* appreciated that the galaxy was not just a bunch of point-like, featureless stars and a whole lot of blackness between. So, why did I pick on Orion? Apart from being my favourite constellation, I want to use it as an illustration that the stars above us are not all of the same type. At first glance, all stars look roughly the same – just points of white light of varying brightness. But this is not true. Find a dark site far away from the glow of houses and street lights and let your eyes adjust to the darkness. You will notice that some of the bright stars are slightly different colours.

With the naked eye you can only see a few thousand stars at any one time (not counting the combined glow of billions of unresolved stars of the galactic disc). The stars we see are generally the ones in our solar neighbourhood – so just the tip of the iceberg. Or, to drill home a point, like a hydrogen atom in a water molecule at the tip of an iceberg which is floating in an ocean. What connects me to the universe is the idea that, although the cosmos is like a chasm, most of which is beyond the scope of my human senses, through the science of astronomy I can dramatically extend my senses, seeing further than my biological limitations allow.

Many different stars

The brightest star in the top left (shoulder) of Orion is Betelgeuse, and is called a red supergiant. Look carefully and you will notice it has a reddish colour. Betelgeuse is a young, massive star, around twenty times the mass of the Sun. It's a youthful 10 million years old (the Sun is geriatric in comparison, having shone for around 5 billion years), yet Betelgeuse is already in its final days.

The more massive a star is, the faster it 'evolves'. What we mean by that is the rate at which the star consumes its fuel supply of hydrogen gas. Once the gas is used up, the star begins to die, because it can no longer stay in equilibrium with the force of gravity, a force that is always trying to collapse and crush the star, and the outward pressure generated by the energy released by nuclear reactions in its core. All stars experience this battle throughout their lifetimes, but because the gas supply is finite and gravity is patient, all stars must come to an end. The nature of that ending depends on the mass of the

The Horsehead nebula, a rearing filigree of dense gas and dust backlit by light from the emission of ionized gas in the Orion nebula. Optical wavelengths of light cannot penetrate this structure, causing the Horsehead to stand out clearly against the background.

star. In the last chapter we talked about supernovae, the very violent deaths of some stars. Aside from the white dwarf systems that can accrete matter from a companion star in a binary orbit and explode (the Type 1a supernovae), only stars about ten times as massive as the Sun can explode as supernovae on their own, when the collapsing core – no longer 'held up' by radiation pressure – exceeds the mass and density necessary for a runaway thermonuclear reaction. We call these Type 2 supernovae. Betelgeuse is one of those massive stars destined to explode as a Type 2. When it eventually does go bang (it's already shedding material in an outflow from its surface, which is the astronomical prelude for a stellar crescendo) it will be bright enough to see during the day. As well as being more massive, Betelgeuse is also physically much larger than the Sun. If it were placed at the position of the Sun, Betelgeuse would engulf and incinerate everything within the solar system out to beyond the orbit of Mars. That's one single star that is physically the size of the entire inner solar system. Quite remarkable.

Look diagonally down the constellation to the foot of Orion and you will see another bright star, again a giant: Rigel. Rigel is a *blue* supergiant, and should appear blue-white if you let your eyes become accustomed. It is nearly 100,000 times as luminous as the Sun, and, at its distance of about 260 parsecs from us, appears as one of the brightest stars in the sky – beautiful on a crisp winter's night. Scan around all the other stars in the sky and you should see a mixture that varies between red, blue and yellow.

What is the origin of the different colours of stars? As we have seen, in astronomy, we measure the colour of something as the difference in brightness through two different photometric filters, or, more generally, at two different wavelengths of light. Combined with a measurement of the star's luminosity, the stellar colour provides a means to classify all stars. The colour of a star is a reflection of its surface temperature: blue is hot, red is cool. The principle behind this is exactly the same as what happens when you take, say, a metal bar and heat it with a blowtorch. It will first glow red, then orange, and progress to blue-white as it gets hotter and hotter. The hotter the bar is heated, the more thermal energy it builds up. This thermal energy is radiated as light – electro-magnetic radiation – and the exact temperature sets the frequency, or colour, of the light being emitted. The same principle applies to stars. As a guide, the surface temperature of the Sun is something like 6,000 degrees Celsius.

If we make a chart that plots stars' luminosity against their colour, we find that the points are not distributed haphazardly: there is a fairly tight locus along which most of the stars lie. The bluest or hottest stars are the most luminous and the reddest or coolest stars the least. This plot is called the 'Hertzsprung-Russell' diagram after the astronomers Ejnar Hertzsprung and Henry Norris Russell, who pioneered it in the early twentieth century. The locus where most of the stars lie is called the 'main sequence', and the location of a given star on the main sequence is determined by its mass. Along the sequence, there is a continuous range of star temperatures, and we break this range into bins, or classifications, called 'spectral types'. The *exact* spectral type is defined by the chemistry of the star as revealed by its spectrum, which shows a range of emission (hydrogen and helium) and absorption (metals) features. However, as a first-order approximation, the colour of a star is a good proxy for its surface temperature and therefore its spectral type. Since the surface temperature and luminosity of a star are both related to its mass (in slightly different ways), the position of a star on the main sequence provides us with a method of estimating its mass.

The spectral types are coded (from blue/hot to red/cool) as O, B, A, F, G, K and M. Modern classifications break up this sequence further into a finer range of types, but that's not important for now. As a guide, the Sun is a G-type star. Rigel is a B-class star and Betelgeuse is an M-type star. The Sun is a 'main sequence star', but Rigel and Betelgeuse do not lie on the main sequence but on other loci, called the 'giant branches'. Stars only reside on the main sequence while they are burning hydrogen. After the hydrogen is used up, different nuclear reactions start to occur within a star, and it 'evolves' off the main sequence. For a star like our Sun, the start of this evolution happens as hydrogen is used up in the core, which becomes saturated by helium (which is one of the products of hydrogen burning). As this happens, the size of the region where nuclear reactions take place will expand as the remaining hydrogen in the star is burned in 'shells', rather than a central core. The move from 'core' to 'shell' burning means that the Sun will physically expand into a red giant, as energy output starts to skyrocket due to the increasing pressure in the core, causing an increase in luminosity that will incinerate the inner solar system. At the end of its life, when no more nuclear reactions can be sustained in the remaining gas, the Sun will increase in temperature and eject its atmosphere (which, remember, has been enriched with new elements formed during the fusion occurring throughout its lifetime), leaving a cool, compact remnant – called a white dwarf – at the centre of a so-called 'planetary' nebula, the expanding, diffuse remains of what was once star stuff. The most massive stars, like Rigel and Betelgeuse, evolve quickly, giving rise to the giant branches that extend away from the tip of the main sequence. Since they die so quickly, these stars are often found close to the sites where new stars are actively being formed (like Orion). It is the ultraviolet and blue light from O and B stars that contributes to actively star-forming galaxies' blue colours.

It's important to remember that, while on the main sequence, all stars emit the bulk of their energy in the ultraviolet and visible light bands, although where the peak of this emission is depends on the spectral class, and this corresponds with the colour of the star. Intense ultraviolet radiation – as we know from sunburn – is harmful to biological systems, and the Earth's atmosphere blocks out most of it. Most of the optical photons do get through, and it's no accident that human and indeed most animal eyes, as well as plant biology, have evolved to be sensitive to this radiation. So, as well as the throwaway line often bandied around that 'we are stardust', the fact we see the world in a narrow range of optical light and not ultraviolet or infrared or some other wavelength is another link, and everyday reminder, of the connection between biological life on Earth and the physics of stars.

Thus high-mass stars live for a short amount of time, but some of the lowest mass stars have the potential to live for trillions of years, much older than the age of the universe, which current measurements indicate is just under 14 billion years old. As the universe continues to expand and galaxy evolution progresses, it will be the lowest mass stars that will be left behind as more massive ones start progressively dying off. Stars will continue to form in galaxies as long as there is atomic fuel that can collapse into dense molecular clouds, but one day even those reservoirs will be exhausted. What will be left is a huge universe sparsely populated with ghosts of galaxies: dim, red systems containing an ever-ageing population of ancient stars. Then one day, far in the future, everything will go dark. But happily, now, and for billions of years to come, the universe is bright.

Galaxies contain stars of different sizes, masses and ages, and we can distinguish them observationally using their colours and luminosities. But *why* is there a mixture of stellar types in our galaxy? Why aren't all stars formed equal? First, forget about the whole galaxy; let's just consider a single star-forming region. Stars form within large clouds of gas, which form from gravitational attraction: more tenuous atomic hydrogen can condense to form clouds of molecular hydrogen, which in turn can collapse further under the force of gravity, not into one single point, but into many dense regions, fragmenting into many cores. This is because the cloud is not smooth; there are regions within it that are denser than others. This is a natural result of the formation of the cloud and the turbulent motions within it.

The densest regions in the cloud tend to collapse first. Initially, the dense clumps form 'proto-stellar' cores – dense globs of gas suitable for the potential ignition of nuclear fusion. If the density of the core reaches a sufficient level, nuclear reactions can take place, and a star is born. So, a single cloud of gas doesn't collapse into just one star – it can form multiple stars in a given generation, creating a star cluster. When they are formed, stars can be in motion, because they are imparted with some of the angular momentum of the collapsing regions from which they formed. This means that the new stars start to drift away, migrating out of the birth ground like chicks from a nest. Most importantly, there is also a range in the masses of these collapsing cores – just as humans are born within a range of weights, so there is a particular distribution in the masses of the natal stars. The form of this distribution is called the 'initial mass function', or IMF. The IMF simply describes, for a given generation of stars, how many stars there are of different mass. Alternatively, it can be viewed as a probability distribution: if I pick a new star at random, where is it likely to be on the main sequence?

We know, when we look at specific regions within our galaxy, that if we see clusters of massive, young stars, we have found an actively star-forming region, because these stars must have formed fairly recently and have not yet died off. We can play the same trick with distant galaxies to identify those that are actively forming stars. Determining the shape of the IMF is also vital in our interpretation of galaxies because it allows us to estimate the total stellar mass of a galaxy. Remember, we don't actually measure mass directly; we measure light, usually the blended emission of all of the stars in a galaxy. The IMF allows us to translate the total stellar luminosity of a galaxy to a total stellar mass. This is analogous to calculating the total weight of a group of people just by counting them, provided you have a previous estimate of what the typical distribution of human weights is.

It is still not clear if the IMF is universal and unchanging over time, and indeed the exact origin of the shape of the local IMF is still in some debate, and this is one of the current major uncertainties in galaxy evolution studies. I don't

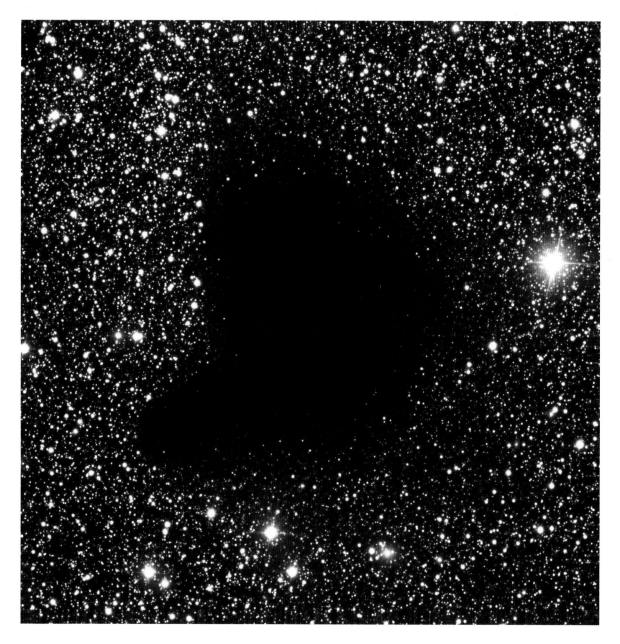

Obscuring dark cloud Barnard 68, one of Edward Emerson Barnard's catalogue of 'dark nebulae' in the Milky Way. This cloud of dense molecular gas and dust is about twice the mass of the Sun, and is close enough to us that there are no intervening stars. The cloud is opaque to the light of background stars, making it appear totally dark in optical images like this. Barnard 68 will gravitationally collapse in the future, possibly giving birth to a new star.

Another image of the dark cloud Barnard 68, this time incorporating visible *and* near-infrared light, which is encoded as the 'red' channel in this composite image. In contrast to the image of Barnard 68 taken in visible light only, where the cloud appears totally opaque, in this image we can see stars through the cloud: this is a great example of how near-infrared light can pass through the dense, dusty and gas-rich medium more easily than shorter-wavelength visible light, which is readily absorbed.

The Lupus 3 dark cloud: a dusty molecular region within our own galaxy that is opaque to visible light such that it obscures the view of the stellar background field. The bright blue stars are new, young, hot stars that have just formed and emerged from the dense dark cloud. The blue haze surrounding them is the starlight scattered off dust and gas in the immediate stellar environment, like the haze surrounding a street lamp on a foggy night.

The Taurus molecular cloud is a large complex of dense gas and dust in our galaxy; another place where new stars can form. In visible light, this part of the cloud appears as a dark band across the stellar field, because the dense gas and dust block out the light of background stars, and the cloud itself does not emit visible radiation. But this image combines observations from a telescope that operates in the sub-millimetre part of the electromagnetic spectrum. Cold dust (at temperatures of a few tens of degrees above absolute zero) emits 'thermal' radiation at far-infrared and submillimetre wavelengths. So this dark tendril of gas and dust glows brightly when imaged at these longer wavelengths. This composite optical and submillimetre image shows in orange the thermal emission of the molecular cloud, revealing dense, bright nodes where new stars are poised to pop into life. Some distant, very actively star-forming galaxies are so dusty that most of their emission emerges in the far-infrared and submillimetre bands, and are in fact virtually invisible in visible light; we rely on infrared and submillimetre observations to detect them.

The Tarantula nebula, or 30 Doradus, a star-forming region located in the Milky Way's satellite companion the Large Magellanic Cloud. Environments like this are referred to as HII regions, owing to the emission from ionized atomic hydrogen.

A close-up view of 30 Doradus. This region is alight with the firework-like cluster of blue stars. These very massive and hot stars have recently been formed and emit copious amounts of ultraviolet and optical radiation, evacuating a bubble in the nebula and causing the surrounding gas to glow as the gas becomes ionized. This image shows the light of ionized oxygen and hydrogen.

want to say too much about stellar evolution. That's a book in itself. What I hope I have conveyed here is the fact that there are different types of stars in the galaxy, living for different lengths of time and emitting different amounts of energy.

Gas: The blueprint of star formation

We explored a hypothetical individual gas cloud above, which collapses gravitationally and churns out new stars. That was just an illustration of the way stars are formed. What is the distribution of stars and gas in the Milky Way as a whole?

As we know, the Milky Way can be divided up into its disc and central bulge – the white and the yolk, if you like. The disc of the galaxy is where most of the dense gas reservoirs responsible for forming new stars are located, and these are so-called Giant Molecular Clouds (GMCs). They are called 'giant' because they are large, spanning some hundred or so parsecs, and contain enough fuel to form (potentially) millions of new stars. They are 'molecular' because the gas within them is primarily composed of molecular hydrogen, the simplest molecule: just two protons bound together by shared electrons forming a simple covalent bond. In order to form in the first place, molecular clouds must have 'cooled' from more tenuous gas where the hydrogen atoms were not yet bound together. We say the gas has 'cooled' because, for the molecules to form, those atoms must get close enough together so that they became bound via the electromagnetic force, and don't simply zip by each other. This is the situation in hot gas: the atoms have lots of energy, corresponding to high velocities. This energy must be lost, or at least reduced, if molecules (and subsequently, stars) are to form.

At first it's a bit confusing to think that stars, which are hot, form from gas that has cooled, but what we really mean is that the gas cloud as a whole has collapsed gravitationally, losing some of its internal energy so that fusion – star formation – can eventually take place in dynamically cold clumps. Once stars start forming within a cloud, the gas around the sites of new star formation starts getting blasted by the radiation and winds driven by those new stars. This backlash not only ionizes the surrounding gas, creating a glowing nebula (like Orion), but the combination of the radiation and winds blown by the stars starts to blow out bubbles and cavities within the GMC, affecting the distribution and chemistry of the gas. Thus the astrophysics

The VISTA telescope's near-infrared view of star formation in a region called Monoceros, a region of active star formation in the Milky Way. This is part of a giant molecular cloud, but here we see only the parts of the cloud where new stars – the majority of which are clustered in a dense concentration near the centre of this image – are illuminating the surrounding gas and dust, which reflects and scatters that light. The wide eyes of VISTA are perfect for capturing panoramic images of huge environments such as this, which can span several degrees on the sky.

A remarkable view of a site of star formation in the Milky Way – the Carina nebula – taken with the Hubble Space Telescope. This image shows the nebulous glow of hydrogen, sulphur and oxygen energized by the light emitted by newly formed stars, which are also shaping the nebula as radiation and winds driven from the stars' surfaces erode and carve the surrounding gas and dust clouds.

The most massive stars formed in this nebula will die quickly, within a few million years of their birth, and their deaths in supernova explosions will inject energy and new elements back into the interstellar medium. The elements sulphur and oxygen, which contribute to this kaleidoscopic vista, are testament to this: these elements were formed in previous generations of stars.

at the interface of star formation and the interstellar medium is incredibly complex, meriting a dedicated field of astrophysics research.

There are many GMCs spread throughout the galactic disc. If we could view the Milky Way from above, we'd see many patches of red-hued ionized hydrogen and clusters of blue, young stars punctuating the spiral arms of the galaxy. We cannot get to this vantage point for obvious reasons, but images of nearby spiral galaxies that present their faces to us give us an excellent idea of what the Milky Way looks like from the outside.

We measure the star-formation rate – or SFR – of a galaxy in the convenient units of the equivalent mass in Suns formed per year. The Milky Way has a star-formation rate of just a few solar masses per year, and it's important to consider that even after billions of years of evolution, the galaxy has not yet used up all of its gas: it remains an active place, albeit comparatively sedate compared to the most extreme galaxies in the universe, which we'll come to. If we waited long enough and watched the evolution of the Milky Way, pretty much all the gas in the galaxy would be turned into stars, and the supply of gas from the surrounding intergalactic space – which gradually rains down via gravity – would turn into an insignificant trickle.

A few tens to 100 million years later, after the last generation of stars forms, the massive stars would die, leaving behind their longer-lived but less massive cousins. The disc would eventually fade and turn from blue to red as the bluer spectral types die off progressively. Such galaxies do exist, and are called 'passive spirals'. They are thought to be typical spirals in which star formation has ceased, either because of some environmental influence that prevents gas from forming new stars, or because they have run out of fuel.

Detail of the birth of stars in the Carina nebula, focused on a 'pillar' of gas and dust within which stars are forming. This is just part of a larger complex of star formation within a huge cloud of gas; a scenario played out in patches throughout the disc of our galaxy and other star-forming galaxies in the universe, wherever there is a reservoir of cold, dense gas and the conditions are right for the thermonuclear triggering of star formation. The pillar is quite opaque, even to the intense light emitted by the new stars within it, but jets emitted by some young, massive stars within the pillar can be seen blasting laterally out of the column, and the whole region glows with the light of ionized gases and scattered light. Star formation is an energetic process: radiation and winds from the most massive, young stars can dramatically alter and shape their immediate surroundings, and form part of the feedback energy responsible for regulating the growth of galaxies.

On the other hand, if the Milky Way collides with another galaxy, as it will probably do with M31 in the future, there will be a violent event that could significantly boost the star-formation rate. Strong gravitational tidal forces will distort and tear the two galactic discs, triggering bursts of star formation in disturbed clouds, which are impelled to collapse from the gravitational perturbation. No stars will physically collide – they are so small and far between that the chances of individual stellar collisions when galaxies collide

Two views of the same galaxy. The left panel is an image of the galaxy M83 in near-infrared light. On the right is an image in visible light. The pink and blue hues in the visible light image show new stars and the ionized gas of HII regions, mostly in the disc and spiral arms of the galaxy. These are invisible in the near-infrared image, because the young, massive stars that emit most of their energy in the ultraviolet and blue part of the spectrum do not contribute much near-infrared light. Conversely, the older, more mature stellar population in the galaxy contributes a lot of near-infrared light, so the central bar and bulge are more prominent in the left-hand image, although you can also see the clusters of red giant stars that are associated with the star-forming regions visible on the right. We do not see such prevalent dust lanes in the near-infrared image because these photons can cut through the interstellar dust more easily than bluer photons, which are easily absorbed. With these complementary views of galaxies at different wavelengths of light we can peel back different layers, almost in an act of dissection, learning about the many different components of galaxies and the links between them.

The Flame nebula, a star-forming region in the Milky Way close to the Orion nebula, again imaged in the near-infrared part of the spectrum. This allows us to see through much of the interstellar dust that blocks out light at bluer wavelengths, revealing the bright young stars forming within this dense environment of gas and dust, illuminating the 'walls' of the nebula around them.

are very low. We see these starbursts happening in other galaxies that have recently collided; stellar discs are ripped into long tails, and there are patches of intense ultraviolet and infrared emission, often towards the dense centre of the system. When things settle down, our galaxy will have changed chemically, dynamically and structurally. New generations of stars and the new solar systems that form with them will be enriched with elements that will literally have formed a long time ago in a galaxy far, far away.

Galaxy collisions are events that stir things up: they deliver (or bring together) new material and promote new growth. As always, the dense gas is where all the action happens, but this gas is surprisingly difficult to detect. Most of the molecular hydrogen in galaxies cannot be observed directly, because – for physical reasons related to the structure of the hydrogen molecules – under normal conditions it doesn't emit radiation that we can detect. And yet molecular hydrogen is a fundamental component of galaxies, so how can we learn about the properties of the raw material for star formation?

It's easy to see the glowing, ionized gas around star-forming regions, but these are like brushfires in a more expansive savannah. The majority of the gas in any one GMC is not actively forming stars. So, how do we measure and map the molecular gas? The answer comes from the contamination of that gas by previous generations of stars. One of the most common molecules in galaxies after hydrogen is carbon monoxide. This is the same stuff that is emitted by poorly burning gas fires, and which you can detect in your home. Carbon monoxide tends to be mixed in with the hydrogen gas, which is extremely useful because, unlike the hydrogen molecules, it *does* emit radiation when excited into an energetic state. In this case, that energy is in the form of the simple rotation of the carbon monoxide molecules (which are single carbon and oxygen atoms bound together). This rotation can happen when carbon monoxide molecules collide with hydrogen molecules. As we have discussed, changes in the energy of quantum systems (like molecules) result in the emission of precisely tuned radiation. At the molecular level, even the rotation of a molecule like carbon monoxide is regulated by quantum mechanics: only certain types of rotation are allowed. This means that carbon monoxide, when rotationally excited, emits radiation at regular intervals in frequency. Different frequencies of emission correspond with different energy states; the highest frequencies are emitted by carbon monoxide molecules in the most energetic states and vice versa. These energy states are dependent on the density and temperature of the gas.

It takes gas densities of a few hundred particles per cubic centimetre and temperatures of a few tens of degrees above absolute zero to start emitting the lowest-energy carbon monoxide lines. In context, the gas that is producing this emission is representative of the bulk molecular gas reservoir. Unlike the emission lines of ionized gas we have talked about in the visible light part of the spectrum, the carbon monoxide emission has wavelengths of the order of a millimetre, between the far-infrared and radio part of the spectrum, so it cannot be observed with a normal optical telescope. Instead we can use radio (or, more precisely, millimetre-wave) telescopes equipped with suitable receivers that can detect photons of this wavelength. Once we detect the carbon monoxide emission, we can measure the total amount of light and convert this to carbon monoxide luminosity (assuming we have some estimate of how far away the emitting gas is). Since the carbon monoxide emitting gas is mixed in with the molecular hydrogen such that the more hydrogen there is, the more carbon monoxide there is, we can convert the observed carbon monoxide luminosity to a molecular hydrogen mass. Thus, we can tell how much gas is available for star formation in a GMC, or indeed a whole galaxy.

Traditionally this has been quite a challenging observation for galaxies much beyond our local volume – the technology has not been available to detect the faint carbon monoxide emissions from very distant galaxies (apart from the most extreme, luminous galaxies like quasars). All this is changing right now with the development of a new telescope – or rather, an array of telescopes – called the Atacama Large Millimetre Array (ALMA).

ALMA is a collection of about 50 radio dishes, each 12 metres in diameter, spread over a large area of land in the high Chilean Atacama on the Chajnantor plateau, at an altitude of about 5 kilometres. ALMA is an international project, with major contributions from the United States, Europe and Japan. The magical thing about an array of telescopes like ALMA is that they can be linked together electronically to act like one very large telescope, utilizing the light-collecting area (remember the idea of light 'buckets') of all of the dishes and attaining very high spatial resolutions. This technique is called interferometry. ALMA is incredibly sensitive in the sub-millimetre and millimetre bands and, once it reaches full operational power, will be able to detect the molecular gas in galaxies not dissimilar to the Milky Way, but seen close to the start of cosmic time. It's an amazing leap forward in this area of astronomy, and is ushering in a new era of exploration of galaxies that will yield fascinating discoveries for several decades to come.

The Atacama Large Millimetre Array with the Magellanic Clouds seen as the two fuzzy clouds of light above, among the foreground stars within the main disc of the Milky Way itself. The Clouds are two large dwarf companion, or satellite, galaxies to the Milky Way. Most large galaxies have an entourage of satellite galaxies, but predicting the number and distribution of these satellites is a current issue in detailed models of galaxy formation.

We've talked about the molecular gas – the building blocks of stars – but it's important to also consider the other major gaseous component of galaxies: the neutral (that is, not electrically charged) atomic hydrogen, HI, which precedes the molecular phase. This HI gas comprises single atoms of hydrogen rather than molecules of hydrogen. Unlike the molecular hydrogen, the atomic component is more diffuse and is not restricted to dense, compact clouds trapped in the disc. The atomic hydrogen is incredibly useful as a tracer of the outer edges of disc galaxies, where the density – and therefore brightness – of stars starts to peter out. Atomic hydrogen is easy to spot because it is a strong emitter of radio waves. Not any old sort of radio waves, mind you – in the rest frame, the gas emits light at a frequency of precisely 1.4 gigahertz, or equivalently a wavelength of 21 centimetres. Like the precise carbon monoxide emission from GMCs discussed before, and like those ionized gas emission lines around star-forming regions we've talked about, the 21-cm emission from atomic hydrogen is also an emission line. This time the physics of the emission is slightly different again. I'll explain it, because it illustrates two important things: one, the ridiculous numbers involved in astrophysics; and two, another nice link between quantum mechanics and astrophysics.

Hydrogen atoms are made from a proton and an electron. In quantum mechanics, these particles have a property called 'spin', which doesn't really have an analogue in classical physics, but is a bit like a quantum angular momentum. Anyway, the spin of the proton and electron can each be thought of as orientated up or down, so it's easy to think of a bunch of hydrogen atoms, some where both the proton and electron have their spins in the same direction (parallel), and some where the spins are in opposite directions (anti-parallel). It turns out that the quantum state in which the spins are parallel has a little more energy than the state in which they are anti-parallel. A quantum system is lazy – it likes to be in the lowest possible energy state – so there is a mechanism by which those atoms with parallel spins can have the electron 'flip' so that its spin points in the opposite direction to the proton's spin. This is called hyperfine splitting, because the difference in the energy between the parallel and anti-parallel states is tiny compared with the overall ground-state energy of a hydrogen atom.

The energy that the system loses in this transition has to go somewhere, so every spin flip releases a photon with a very specific energy corresponding to the exact difference in energy between the parallel and anti-parallel states, which happens to correspond with electromagnetic radiation – a photon –

with a wavelength of precisely 21 centimetres. The corollary is that neutral atomic hydrogen can also absorb radiation with a wavelength of 21 centimetres, where energy is absorbed by the atom and 'stored' by aligning the spins of the electron and proton.

Hyperfine splitting is called a 'forbidden' transition because for any one atom, there is a *very* small chance of it occurring under normal conditions. In fact, the chance is so remote that if you observed a single hydrogen atom aligned in the parallel state and waited for it to undergo the hyperfine transition, you would have to wait on average 10 million years for it to happen. If you observed 10 million atoms, then you would expect to see just one photon released per year. That's still not much of a signal. In astrophysical scenarios, however, we can exploit atomic crowdsourcing: there are so many neutral hydrogen atoms in an astrophysical cloud of gas that the radio emission is really quite bright – since at at any one time, a huge number of 21-cm photons are being emitted via the hyperfine transition. I find this amazing – this is a probabilistic quantum mechanical release of a photon from a single atom that simply doesn't happen on Earth, but when it is put in an astrophysical theatre it gives rise to one of the most important observations we have of our own, and indeed other galaxies. Radio telescopes with receivers that can be tuned to detect 1.4 GHz can map out the atomic hydrogen in our own galaxy and other nearby galaxies.

Again, like the carbon monoxide measurements, the detection of atomic hydrogen much beyond the local volume is difficult. Like all electromagnetic radiation being emitted by a source moving relative to us, the 21-cm line is subject to redshift, which stretches wavelengths longer and equivalently makes frequencies lower. The rest-frame frequency 1.4 GHz is already quite low. Make that lower still and it moves into a part of the radio frequency range that is quite difficult to detect. For one thing, below 1 GHz we get into the radio bands used commercially for TV and

An image of neutral atomic hydrogen gas (HI) in a map of the galaxy NGC 628 from a survey called 'The HI Nearby Galaxy Survey' (THINGS). HI gas emits radio waves with a very specific frequency, 1.4 GHz, so by tuning radio telescopes to this frequency one can map out the neutral atomic gas in galaxies. This image combines ultraviolet and infrared light (pink/purple) with a radio image detecting the HI gas reservoir (cyan). The atomic gas observations extend the spiral structure far beyond the stellar disc (so HI observations can be used to probe the outskirts of gas-rich galaxies), and represent material that can fuel star formation, provided it can collapse down into dense clouds where molecular hydrogen is able to form (two hydrogen atoms bonded together) and star formation can ensue.

This image of the spiral galaxy M83 combines ultraviolet light observations from the GALEX satellite, highlighting the light from young, massive stars (blue/pink spiral), with radio observations of the emission of neutral hydrogen gas (red). Note how the neutral hydrogen extends far beyond the main stellar structure and is traced in parts by blue star clusters – signatures of star formation in the extended gaseous arms. Neutral atomic hydrogen is the building material of galaxies and can be used to trace the outer galactic environment where there are few stars.

radio, and for communication. This manmade radio frequency interference (RFI) dwarfs astronomical signals, making astronomical observations near-impossible at frequencies that coincide with these ranges. Radio telescopes that want to operate close to the frequencies used for communication must be put in locations remote from terrestrial radio sources (such as the remote part of Western Australia) in order to minimize RFI. The Earth's ionosphere also affects the traversal of radio frequencies below 1 GHz in a similar way to how optical light is bent and refracted by a glass of water, and correcting for this is hard. There are numerous other technical reasons why low-frequency radio astronomy is challenging, but many of these hurdles are now being overcome with the development of large antenna arrays coupled with extremely powerful computers that can handle the insane level of signal processing that must be performed in order to distil astronomical signals in the radio part of the spectrum.

One such recent example is LOFAR: the LOw Frequency ARray for radio astronomy. LOFAR is an array of thousands of very cheap antennae (that actually just resemble black slabs, rather than the parabolic dishes that you usually associate with radio telescopes), spread over a 100-kilometre region in the Netherlands, as well as stations up to 1,500 km away in various parts of Europe: Sweden, Germany, Britain and France. The telescope is designed to detect radio frequencies of 10 to 250 MHz – suitable for exploring what has been dubbed the 'low-frequency universe'. What makes LOFAR different from traditional telescopes is the fact that the antennae are omni-directional – they can record the entire sky at once. Then, in order to observe a particular spot on the sky, the signals from all the antennae are collected and the 'aperture' actually defined within software, using a supercomputer that cleverly processes the signal received by each of (or sometimes just a subset of) the antennae. Although it still requires antennae to do the receiving, LOFAR is basically a digital telescope that has only been made possible through modern computing – the power and sophistication (and affordability) of which will only improve over time.

Like ALMA, LOFAR is a fantastically powerful and innovative telescope that is going to help revolutionize twenty-first century astronomy. One of the goals of LOFAR is to detect the 21-cm line of neutral atomic hydrogen close to the epoch when the first stars and galaxies formed, where the HI emission has been redshifted to very low frequencies – this is the final frontier of galaxy evolution studies. LOFAR has a more practical application too: it is also being used as a sensor network that can be applied to geophysics research and agricultural studies.

Galactic dynamics: dancing to the tune of gravity

We have talked much about the content of galaxies, and how we can measure the different components using various instrumentation and observational techniques. But there is another important observable property of galaxies: dynamics. Galaxies are not static; they move relative to each other with cosmic expansion and also local gravitational attraction. There are also motions *within* individual galaxies that we can measure. For galaxies like the Milky Way, perhaps the most important motion is the rotation of the disc. Our solar system is situated about two-thirds of the way out from the galactic hub, and is orbiting it at a velocity of over 200 kilometres per second. At this rate it takes nearly a quarter of a billion years to make a full galactic orbit (a galactic 'year', if you like). The orbital velocity of our solar system around the hub is controlled by some relatively simple physics; in fact, it is basically the same physics that governs the motions of the planets around the Sun. Simply put, the speed of rotation depends on how much gravitating mass there is between us and the centre of our orbit. In other words, we need to establish how much mass there is in the Milky Way from the centre of the bulge to the radius of the Sun.

When looking at our solar system as an isolated system, the situation is pretty straightforward, because the majority of the total mass of the solar system is in a single point: the Sun. The shapes of the planets' orbits are governed mostly by the gravitational tug of the Sun, and to a lesser extent their mutual attraction. The inner planets rotate around the Sun faster than the outer planets. The distribution of mass in the galaxy is a bit more complicated, but the principle is the same: the speed of rotation of the disc at different radii is linked to the amount of intervening, gravitating mass to the central hub.

Imagine a small sphere centred on the bulge of our galaxy. Now imagine that we can directly add up all of the mass contained within the sphere, and then see how this increases as we expand the size of the sphere, gradually encompassing more and more of the galaxy. Think of it as a galactic mass audit. In practice, remember, when observing galaxies, we can only add up the mass that we can actually see in the form of emitted light. At first glance, it looks as though there is a lot of mass in the centre of the galaxy, with that big, bright, dense stellar bulge. So, as we expand our imaginary sphere, the amount of mass contained within it rises quickly, and we add some more as we move out, taking in the spiral disc. The observed mass stops increasing and levels off once we expand the sphere beyond the disc, where we run out of stars and dust and gas; we have reached the total observed mass of the galaxy. We can do this for other galaxies too (in fact it is easier to do for other galaxies, because we can see the whole thing, whereas for the Milky Way we have the observational inconvenience that we are embedded within the disc). What we just did was add up the mass in the galaxy as a function of radius from the centre. But if the speed of rotation of the disc at different radii depends on the total mass enclosed, then a more elegant method of measuring the total mass of a spiral galaxy is to use its 'rotation curve': the change in orbital velocity of the disc as we move out in radius from the hub.

The physics involved here is old, and will be familiar to any student of classical physics. At play here are the laws of celestial mechanics, first noted by Johannes Kepler in the seventeenth century. Actually it is Kepler's third law, which says that the square of the period of an orbiting body is proportional to the cube of the semi-major axis of its orbit, and inversely proportional to the mass of the attracting body. In other words, for a fixed mass, the larger the radius of the orbit of some body, the slower the speed. Increasing the mass in the system increases the speed of the orbit. Kepler's laws were developed and refined by Isaac Newton, who described orbital motion nearly exactly through the inverse square law of gravity. Newton's description of gravity is not quite right, but this was difficult to see observationally at the time; Einstein's general relativity mopped up the details in the early twentieth century, and that's where our current theory of gravity stands.

To measure the rotation speed, we can turn to the same effect that causes redshift: differences in the velocity of a light-emitting source, relative to some observer (us), result in small changes in the observed wavelength or frequency of the emitted light. So, if the disc is rotating at different speeds, we can track this by measuring the observed frequency of some known emission. Nature has provided us with a convenient way to make this measurement: we can use the 21-cm radio emission from neutral atomic hydrogen gas. Atomic hydrogen is plentiful in galaxies like our own and stretches far out into the outskirts of the disc, making it useful for measuring the rotation right out into the galactic boondocks. The observed shifts in frequency from the rest-frame 21 cm can be turned into shifts in velocity, so if we measure the relative velocities of HI clouds throughout the Milky Way, we can measure the overall rotation of the galaxy. The same thing can be done with the optical light from stars or ionized gas – any emission feature that we can accurately measure a frequency shift in can be used to map velocity this way, it's just that HI is particularly useful as a large-scale tracer here.

You might expect no surprises with the rotation curves of galaxies, as it all sounds fairly straightforward. Actually, when rotation curves were measured properly, they revealed something profound. Astronomers expected the rotation curves of disc galaxies to agree with the Keplerian prediction, assuming the audit of visible matter: the mass given by the rotation curve should agree with the mass that you get when you add up all the stars and gas, and so on. But the data showed something unexpected. If the mass in the galaxy is distributed in the same way as the visible matter, then we would expect the orbital velocity of the disc to increase rapidly out from the centre to a peak, then fall off as we reach the outer edges of the disc. Yet the observed rotation speeds of disc galaxies do *not* fall off with increasing radius; they remain at a fairly steady speed the further out you go from the centre.

Well, this was odd – clearly the theory wasn't matching the observations. Luckily there was a solution, or at least a hypothesis. One way of producing such a flat rotation curve is to include some extra mass, an additional component within galaxies, which is spread throughout the galactic environment in such a way that the mass density towards the outer parts of the disc, where the stars start to become sparse, remains fairly uniform. The flat galaxy-rotation curves are some of the key observational evidence of dark matter. This research was really pioneered by American astronomer Vera Rubin in the late 1970s; Rubin's excellent observational work obtaining rotation curves from precision spectroscopic measurements of spiral galaxies was the first to establish that galaxies like the Milky Way are actually dominated by dark, not normal, matter. Actually, the enigmatic astronomer Fritz Zwicky had already proposed the existence of dark matter in the 1930s to explain the motions of galaxies in massive clusters. However, Zwicky was an antagonistic character, and his hypothesis was largely ignored by the astronomical community. As observational evidence for a dominant dark matter component grew, however, our picture of galaxies changed.

There is some mystique surrounding dark matter, but this is simply because we don't fully understand it. We can see its effects quite clearly in evidence like the rotation curve, but we have not detected this missing matter directly. There is also no strong evidence that dark matter interacts with 'normal' matter (I mean the stuff made from protons, neutrons and electrons) in any other way apart from its gravitational influence. Astronomers don't *want* it to be dark – we desperately want to know what it is – but until we directly detect it (and experiments are attempting to do just that), dark matter remains a theoretical component of our model of how the universe works, albeit quite a successful one; it turns out that our current model of the universe that describes dark matter does a very good job of explaining a wide range of phenomena, and therefore we're quite confident that it exists. It's just elusive.

So, to update our picture, the universe not only contains normal 'baryonic' matter, it also contains dark matter. In the current model, dark matter outweighs the normal matter by about five to one, so there is far more dark matter in the universe than the normal stuff that goes into forming gas, stars and planets.

We tend to describe the normal baryonic matter in galaxies as being distributed in 'haloes' of dark matter. For spirals like the Milky Way, the luminous disc is like the coloured centre of an old-fashioned glass marble. The total mass of the Milky Way including the dark matter halo is something like 100 billion times the mass of the Sun, but the largest dark matter haloes in the universe – the ones that contain the clusters of galaxies – can have total masses more than 1,000 times higher than this. We'll come to how we think dark matter and normal baryonic matter comingled to *form* galaxies later, but before we go on, we should further explore the other types of galaxy that exist.

The many types of galaxy and the cosmic web

Although the Milky Way is a spiral galaxy, when we look at other spiral galaxies there is actually a variety of 'spiral-ness'. For example, there is a range in how tightly 'wound' the arms are around the central bulge, and indeed there is a range in the size and brightness of the central bulge itself. This range can be broken into a set of classifications: Sa, Sb, Sc and Sd, where 'S' stands for spiral and a, b, c and d refer to the range of spiral- and bulge-ness, from tight arms with a big bulge (Sa) to poorly defined, clumpy arms with not much of a bulge (Sd).

About 60 to 70 per cent of spiral galaxies have another interesting morphological feature: a stellar 'bar' emanating from the bulge, like a spoke that connects the inner edges of spiral arms. Like spiral galaxies without bars, barred spirals also have a classification: SBa, SBb and so on. Barred spirals are pretty common; actually, it is thought that the Milky Way itself has a bar. The formation of the bar is a consequence of dynamical instabilities, and is thought to form from density perturbations in the disc. One important feature of the bar is that it has a role in transporting stars and gas towards the bulge, potentially fuelling star formation and black hole growth at the centre of the galaxy, and contributing to the overall evolution of the system.

There are galaxies much smaller than the Milky Way; lower-mass systems with no particular shape to their stellar distribution – they are amorphous, irregular systems. We call these dwarf galaxies. They are often forming new stars, but at relatively low rates, and because they contain comparatively few stars, dwarfs tend to be faint, and are therefore hard to spot when at great distances. Dwarfs are commonly associated with a larger galaxy, clinging on gravitationally to the outskirts. The Milky Way has several dwarf satellites, but the most famous (and the biggest) are the Magellanic Clouds we first met in chapter One, which are easily seen from the southern hemisphere.

The fact that large galaxies, like the Milky Way, are accompanied by a retinue of dwarf satellites, and that in turn the Milky Way is part of the Local Group, are clues that the organization of matter in the universe is hierarchical: large-scale structure is assembled from a series of smaller structures. The overall distribution of matter we see was first set down, or encoded, if you like, at the earliest point in the history of the universe – very shortly after the Big Bang – and has been governed by gravity ever since. Large surveys have revealed colossal structures in the distribution of galaxies – even bigger than clusters – called 'walls' and 'sheets' (one of the most famous being the Sloan 'Great Wall', discovered in the Sloan Digital Sky Survey). The flip side is that, if most of the galaxies are arranged into a large-scale filamentary structure, then in the gaps between these filaments and sheets there are vast 'voids' – chasms of totally empty space that are millions of parsecs wide.

At the densest points in this cosmic web we find the clusters of galaxies, like Virgo, that we first met in the last chapter. These clusters are the habitat of the most massive galaxies in the universe, the ellipticals (E). Elliptical galaxies are both physically larger and 100 times more massive than our galaxy. As the name suggests, ellipticals are not flat discs, but bulbous assemblies of stars. Think of a football and then compress it into a rugby ball; ellipticals occur in a range of shapes between these extremes (the ones that are close to football-shaped are often called spheroidals). We classify ellipticals according to how 'squashed' they are (in the parlance, they are 'oblate spheroids'). Elliptical galaxies have a further key distinction from spirals: they are no longer forming new stars and contain little gas relative to the mass in stars. We call them 'passive' galaxies.

Morphologically, ellipticals are pretty dull, with their featureless and smooth stellar distributions. Occasionally we see ellipticals that contain 'lanes' of thick interstellar dust, blocking out the light. This is the residue from an earlier, active phase in the galaxy's life; the detritus of star formation. Not only are they not forming new stars but ellipticals are also very old, and this is evident in their colour: combined, the light from all of their stars is in the red part of the visible spectrum. As we discussed before, this is a sign that all of the young, massive stars that would have been formed in recent star formation have long since died. What's left is a huge population of lower-mass stars that is simply evolving as per the standard track of stellar evolution, and as this proceeds, the galaxy takes on the rusty hue of old age. When we estimate the age of ellipticals (that is, the average age of the stars), we find that the majority of the stars formed very early in the universe's history, some 10 to 12 billion years ago. This is a hint that the ancient universe was a more active place, in terms of galaxy growth, than it is today.

What about the dynamics of ellipticals; how do they compare with the Milky Way? The stars in elliptical galaxies are not distributed in a disc, and they don't move around the core in nice, orderly, circular orbits. Instead they buzz around on radial orbits like millions of comets around a bright focus. Again, the stars' motions are determined gravitationally by the total mass of the system, and as with the rotation curves of spiral galaxies, we can use observational techniques to measure these motions and hence determine the total mass.

The ellipticals are passive galaxies, so they do not usually have bright emission lines in their spectra that we can use to track Doppler changes in frequency caused by bulk motion. What they do have is lots of absorption lines – the little chunks taken out of the continuum light of the stars because of the presence of heavy elements that absorb energy of specific frequencies. In elliptical galaxies, due to the highly evolved nature of the stellar population, there are lots of metals. Like emission lines, absorption lines occur at very precise frequencies; if all of the stars in a galaxy were at rest with respect to each other, then combined, the galaxy's spectrum would show a series of very narrow absorption lines corresponding with each of the elements present in all of the stars. But the stars are not at rest with respect to each other; they are all moving on random orbits, impelled by the gravitational potential of the galaxy. So, instead of all occurring at the same place in the spectrum, the absorption lines of each star are shifted a little bit in frequency, relative to the average redshift of the system. When we measure the spectrum, we can identify each absorption line easily enough (say, a magnesium absorption line), but the width of the absorption is broadened a bit compared with what we would get from a single star.

This broadening is caused by the distribution of the relative velocities of the stars contributing to the spectrum. If we use a spectrograph of sufficient resolution, we can measure the width of the absorption line (in frequency), and estimate the velocity 'dispersion'. Since the velocity width is directly related to the total amount of mass in the system (again, it's Newtonian physics), we have a method of weighing elliptical galaxies (or indeed any 'dispersion dominated' system). This is pretty amazing. Of course, this technique is relatively easy for nearby things, where the spectra can be obtained with very high signal-to-noise, but it gets much more difficult for distant galaxies – far harder than measuring an emission line because in this case we're looking for the absence of light in a particular part of the spectrum, rather than a bright spike that stands out on top of it.

Finally, there is a class of galaxy that appears to be – morphologically speaking at least – some way between ellipticals and spirals. These are called lenticular, or 'So', galaxies (pronounced *ess-zero*), and these also tend to live in clusters, although they can also be found in what we refer to as the 'field' – regions of average galaxy density outside clusters. Like spirals, Sos have a somewhat flattened stellar disc but no spiral arms (hence the class So: Spiral galaxy + zero arms). The stars are fairly smoothly distributed and, like ellipticals, lenticulars are generally 'passive'. Also like spirals, Sos have a stellar bulge at the centre, but this bulge is much larger than in a typical spiral, dominating the galaxy. Because of the smoothness of the stellar distribution and the uniform colour of the old stellar population, it is very difficult to distinguish an So that is orientated face-on to us from an elliptical galaxy, but when the So is tilted a bit so that its disc is seen slightly edge-on, then the difference is clear. A classic example is the Spindle Galaxy seen towards the constellation Draco, which is a lenticular galaxy seen nearly edge-on. The Spindle also has a striking lane of dust, the remnants of previous stellar evolution in the dense disc, appearing as a narrow, dark band stretching across the galaxy, blocking out the light from the stars behind.

Edwin Hubble came up with a classification scheme for elliptical, spiral and lenticular galaxies based on their morphological type (Sa, Sb, E and so on) that was loosely rooted in the idea that galaxies transform into the different types along an evolutionary sequence. We now know that this is not the case, at least not in the way it was originally proposed. The scheme labels spiral galaxies according to their spiral-ness, elliptical galaxies based on their elliptical-ness, and places them along what is known as the 'Hubble Sequence'. This sequence starts off with ellipticals that are close to spherical in shape (E0), and moves through the various levels of elliptical-ness (E1 through E7). Then we come to the S0 class, which is a little bit ambiguous, lying (morphologically) somewhere between a true elliptical and a spiral galaxy. After the S0 class the sequence branches into two arms. One arm contains the spirals, Sa, Sb and so on. The other branch contains barred spirals, SBa, SBb and so on. This bifurcation is the reason why the sequence is also known as the Hubble Tuning Fork. It doesn't represent a physical connection between these types, but is nonetheless a convenient means to classify different types of galaxy, and remains in use today. If an astronomer presents some galaxy and says 'here is an Sab (*ess-ay-bee*) galaxy which is very interesting because of such and such', all other astronomers know what type of galaxy is being talked about. You'll notice that dwarf galaxies don't really fit into the scheme, although they're sometimes tacked on to the end of the spiral sequence as 'Irregulars', joining back together the two arms of the Fork. The Hubble Sequence generally includes all the types of evolved *massive* galaxy, but does not describe galaxies that are morphologically perturbed, such as we find in interacting and merging systems, where gravitational forces distort the regular structure. As we will discuss later, interacting systems are very important in the history of the evolution of some galaxies.

The hearts of galaxies

We haven't talked much about the bulge yet. The bulge of a galaxy – the hub of the disc in our case – is the densest galactic environment. The Milky Way's bulge is so dense with stars and dust that, if we look with optical light, we can't see right to the centre. With near-infrared light we can cut through some of the obscuring dust, because less of this light gets absorbed. What we

The orientation of galaxies on the sky is random (although some gravitational effects could serve to correlate the orientation of galaxies in some environments), and so we see galaxies that are edge-on, face-on and every angle in between. This is nicely illustrated by the Leo Triplet, a group of spiral galaxies that happen to lie in the part of the sky that contains the constellation Leo. Many astronomical objects are named after the constellations in which they can be found (like the Andromeda galaxy); even though the galaxies themselves are much more distant than the stars, the easily identified patterns of the constellations provided a convenient way to communicate where on the sky a particular object could be found.

In this deep image of the spiral galaxy NGC 4911 in the Coma cluster of galaxies, a faint, ghostly stellar emission in the far outskirts of the spiral arms can be seen around the bright central spiral, which is tinted blue and pink with the stellar emission of young stars and nebular light of HII regions, scarred by dust lanes. The extended outer arms are distorted, gravitationally disturbed by the influence of a neighbouring galaxy. In the crowded environment of a galaxy cluster, gravitational interactions between galaxies can be common, and this can morphologically alter disc galaxies such as this one as well as modify their star-formation histories by perturbing, or in some cases removing, their gas reservoirs – the building material for future generations of stars.

An elliptical galaxy polluted by dust. NGC 1316 is a member of a relatively nearby cluster of galaxies called Fornax, and it is thought that the dust in this galaxy represents the shredded remains of spiral galaxies that merged to form this system.

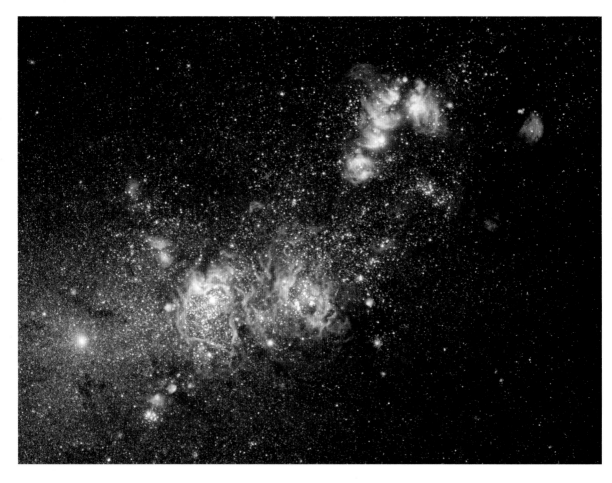

An example of an 'irregular' dwarf galaxy, NGC 4214, undergoing a starburst event. Dwarf galaxies are the lowest mass galaxies in the universe, but often form lots of new stars and usually have no coherent shape. In this view the galaxy is dominated by the light of young blue stars and the glow of ionized hydrogen around the star-forming regions. The new star clusters are actually blowing back the gas clouds they formed from through stellar winds and radiation pressure. This can be seen here: the bright star cluster at lower left is forming a cavity, or bubble, surrounded by HII nebulosity.

NGC 5011B and NGC 5011C: examples of the wide variety of galaxy types. On the one hand we have a bright, edge-on lenticular galaxy glowing with the light of billions of stars, clearly showing a prominent disc and bulge structure, and on the other we have a relatively loose spheroidal collection of bluish stars with a fairly low 'surface brightness'. These two galaxies, although in the same patch of sky, are separated by a huge distance: the dwarf galaxy on the left is fairly close to us, whereas the other is located in the Centaurus cluster of galaxies, some 50 million parsecs away.

This is the irregular dwarf galaxy Barnard's galaxy, in the Milky Way's Local Group. It is a galactic neighbour. Dwarf galaxies are much less massive than the Milky Way, and are often amorphous collections of stars and gas, sometimes clinging to more massive systems. In this image can be seen several red rosettes of glowing hydrogen gas: signatures of actively star-forming regions in the galaxy.

This starry patch is actually a dwarf galaxy called the Fornax Dwarf, a low-mass satellite galaxy of the Milky Way. Although many dwarf galaxies are irregular in shape, this is classed as a spheroidal dwarf, since the stellar morphology is rather circular and regular.

see with a near-infrared eye is myriad stars; the deepest images of the centre of the galaxy reveal almost unbelievable numbers of stars crowded at the heart of the Milky Way. In some ways, the bulge is like a mini elliptical galaxy, containing old stars on random orbits. However, the bulge is not dead; in galaxies like our own, there can still be activity in the central regions of spiral bulges. A common feature of the inner 'nuclear' region of spiral galaxies is a small, dense disc of gas and dust that can form stars at high rates. In some galaxies the rate of star formation in the nucleus can be extremely high because of the high densities of gas that can accumulate there, sometimes transported to the centre of the galaxy by the bar that, as we have seen, is often present in spirals. Although we do believe that the Milky Way is a barred spiral, it isn't terribly extreme in regard to its nuclear activity. But its core is still an interesting astrophysical environment.

Every galaxy with a significant stellar bulge also harbours a compact supermassive black hole at its centre. We learned in the last chapter that in quasars and certain other galaxies, this black hole is actively growing by accreting matter, causing the release of a tremendous amount of energy that can dominate the energy output of the galaxy. In most galaxies, however, the supermassive black hole just sits there quietly in a dormant, quiescent state. Elliptical galaxies, despite not forming new stars, can have active central black holes – also powered by the accretion of matter – that can beam energetic particles (like electrons) outwards at velocities close to the speed of light. When these fast-moving particles interact with other gas and magnetic fields in the galaxy, then radio waves are emitted. We sometimes see a radio 'jet' – a narrow, powerful, collimated beam – that can burst out of the galaxy and into extragalactic space in a spectacular way.

My favourite example is the local Cen A that we met in our local 1-metre cube model. Look at an image of this galaxy taken with optical light, and the galaxy seems like a reasonably normal elliptical (well, it has got a pretty impressive lane of dust across it). But take an image with a radio telescope and you get a very different picture. You don't see the stars (they are not strong radio emitters), but instead see two jets emanating from the core and bursting out of the galaxy, flowering into giant lobes of radio emission when they get into the low-pressure environment of intergalactic space. X-ray observations with the Chandra observatory also reveal high-energy emission associated with the jets, in particular highlighting the hot gas that is 'shocked' to a high temperature as the jets ram into the interstellar and circumgalactic mediums. Without a view of Cen A at other wavelengths of light, in particular radio waves, we would have missed the remarkable astrophysics associated with the central supermassive black hole.

Large area surveys of the sky at radio frequencies are quite common, and reveal a totally different view of the universe. Galaxies that emit jets like Cen A are not uncommon (it's just that Cen A is pretty local and very easy to get detailed pictures of), and many other galaxies can be detected by their radio emissions. Aside from galaxies with active nuclei, galaxies that are very actively forming stars tend also to be strong emitters of radio waves. Again, the cause of the radio emissions is the acceleration of charged particles and their interaction with magnetic fields (which all galaxies are laced with). In the case of these star-forming galaxies, the acceleration comes not from a black hole but from the explosion of supernovae, which drive electrons at high speed, accelerating to velocities up to an appreciable fraction of the speed of light, through the interstellar medium. The electrons from supernovae remnants then encounter and spiral around the magnetic fields in the galaxy, releasing a form of continuum radiation called synchrotron emission as they do so. Thus by detecting radio-emitting galaxies, we are detecting actively growing galaxies. This is often favourable, because radio emission is not susceptible to interstellar dust absorption as is visible light, so we might be able to find populations of active but dust-obscured galaxies that would not be detected in a traditional optical survey.

Some supermassive black holes in the centres of galaxies are currently active, some are not, but they all had to form somehow, so must all have at some time gone through phases of growth. When astronomers started looking at the properties of these black holes and how they relate to the 'host' galaxy they reside in, an interesting correlation was found. It hinted at a fascinating piece of the puzzle of how galaxies are assembled. When astronomers plotted the mass of the central black hole, which can be measured spectroscopically with another velocity dispersion technique against the mass in stars of the surrounding bulge, a clear trend emerged. The bigger the bulge, the bigger the black hole.

To some extent this is not surprising: 'big things are big' is a phrase often used in astronomy. Why is this correlation interesting? What's surprising here is the different physical scales involved. The supermassive black hole and its sphere of influence is millions of times smaller than the size of the surrounding bulge, like a fly in a cathedral. To put it flippantly, how does the central

Radio observations can reveal remarkable things. In this composite radio (purple) and optical light image, the radio galaxy Hercules A is revealed in all its glory. The central elliptical galaxy contains a supermassive black hole, which is 'active' – it is feeding by accreting new matter (gas and dust and stars). This activity gives rise to powerful jets of radio emission, punching out of the galaxy and into extragalactic space. Like a plume of smoke, the radio emission eventually billows out at large galactic distance. Radio sources like this are among the most powerful galaxies in the universe, and are important in the story of galaxy evolution since the energy they deposit in the host galaxy and the local galactic environment can modify the star-formation histories of galaxies in a process called feedback.

black hole 'know' to be big if the bulge is big? If the growth of the central black hole and the growth of the bulge *are* physically linked in some way, what process could control the growth of these two components so that they assemble in tandem? If such a process exists, it must be fundamental to the evolution of galaxies.

A leading theory is that the growth of the central black hole and the stars in the bulge form roughly coevally, and that their growth is linked through a mechanism called feedback. Both stars and central black holes require some material from which to form – that is, fundamentally, gas. Gas collapses to form galaxies and compact objects (GMCs, stars and so on) through the force of gravity, but gravity is not the only force at work in galaxy growth. We have seen how, as a black hole accretes matter, it can result in the ejection of energy in the form of intense electromagnetic radiation and mechanical jets that punch out through the galaxy. This energy does not just disappear into extragalactic space; it interacts with whatever is in its way. Since the central black holes are buried deep in the galaxy, there is a lot of material for the nuclear radiation field and emergent jets and outflows to interact with.

There are two main effects that happen as the central black hole dumps energy into the surrounding medium: the heating of interstellar gas through the passage of shock waves and ionizing radiation, and the actual removal of gas and dust as it is swept up by outflows that emanate from the growing black hole (just look at those jets in Cen A – how can they not have some effect on the interstellar medium as they plough through the galaxy?). What are the consequences? The gas in the galaxy that has been heated by the injection of energy from the growing black hole cannot form new stars because it cannot collapse to form the dense proto-stellar cores required to initiate fusion (it has to lose that additional energy to do so), or, more dramatically, the gas is removed from the local environment entirely. So, the growth of the black hole has had an influence on the growth of the surrounding stars, and therefore regulated the growth of the bulge.

For the same reason, the black hole cannot keep up this onslaught indefinitely – eventually it will remove the material that itself needs in order to grow. Once the black hole ceases to accrete matter, the feedback energy switches off. Wait a while, though, and the surrounding gas will begin to cool and collapse back in (remember, gravity is patient – it is always 'on'). The eventual mass of the black hole and the bulge are linked to the total reservoir of material available, but the interplay between the growth of the black

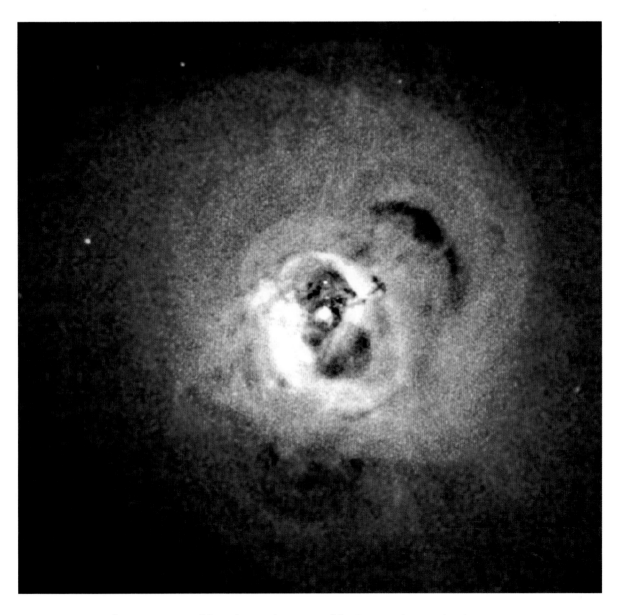

This is an image of the galaxy at the centre of the Perseus cluster taken in x-ray light. The shell-like structures are the signature of feedback from the central galaxy caused by outflows from a central supermassive black hole that has been actively accreting matter. The outflows push back on the hot gas surrounding the galaxy (traced by the x-ray emission), inflating bubbles and shells that propagate outwards. This kind of feedback is responsible for regulating the growth of massive galaxies because it prevents too much gas from gravitationally accreting onto the central galaxy, thus limiting its stellar mass.

hole and the formation of new stars surrounding it is a form of regulation that is thought to result in the observed correlation between central black hole and bulge stellar mass. This is an intense area of research both theoretically and observationally. It is difficult to peer into the hearts of galaxies, so our view of the exact astrophysics is unclear. However, as observations improve (finer vision, better sensitivity), we are gradually adding to our understanding of the details of the growth of the central regions of galaxies.

Islands in a sea of gas

The feedback mechanism turns out to be a very important part of our current model of galaxy growth. Computer simulations (primed with models of structure formation) that aim to grow galaxies in toy universes tend to produce far too many very massive galaxies that we don't see in reality if there is no feedback in the model. Essentially, with nothing pushing back against gravity, galaxies grow too big. Feedback provides a natural modulation of galaxy growth. But feedback is a complex area, since it involves a lot of different aspects of astrophysics, many of which are still relatively poorly understood. Much effort today is being spent on observing the feedback process in action. It's not just black holes that are important: *any* energy that gets dumped into the interstellar medium, be it from the detonation of supernovae or the stellar winds and radiation from the stars themselves, plays a part in feedback.

The mechanisms behind feedback don't just affect the rate of galaxy growth; they also have an important role in distributing and mixing the metals produced in stars around the interstellar medium. In some cases, this dispersal can even extend to the ejection of metals out of the galaxy and into extragalactic space. One way to see this in action is to identify a distant galaxy that is near (on the sky) to, but not exactly lined up with, an even more distant quasar. When we take a spectrum of the background quasar and examine it, we can find in that spectrum the absorption lines from metals in extragalactic space associated with the foreground galaxy. A common element used for this is ionized magnesium, which has absorption lines that are detectable in the visible part of the spectrum and are therefore fairly easy to observe.

What is happening here is that the bright light from the distant quasar – which we're simply using as a convenient backlight – has passed through some gas that has been ejected by the foreground galaxy. Some of that quasar light got absorbed, leaving a characteristic imprint on the quasar's spectrum. This is a beautiful illustration of the fact that the space between the galaxies, beyond the stars, is not empty; it also contains the products of stellar evolution, driven out by galactic winds. Some of this material can later rain back down on the galaxy, enriching the disc as it is re-accreted by the disc's gravitational pull. Other gas might never return to the galaxy if it is ejected sufficiently fast not to be 'grabbed' back by the gravitational attraction, like a rocket reaching escape velocity to break from Earth's gravity. What I hope this all points to is a view I'd like you to take away when you read this: that the story of galaxy evolution is really all about the flow of gas, both within galaxies and to and from extragalactic space.

We know that on large scales the galaxies are organized into a network of filaments, groups and clusters, forming a cosmic web. Aside from the enriched gas surrounding the galaxies themselves, which is first processed by stars and then ejected out into space, the entire network also contains lots of gas that formed in the Big Bang but has never formed or been processed in galaxies. The temperature and density of this intergalactic gas varies, but it is generally quite hot compared with the 'cold' gas that has cooled into galactic discs. So, in some ways, galaxies are not islands of light but just the bright tips of an ocean of gas, like white horses on a choppy sea. There are some environments where this intergalactic medium is very obvious – like the clusters of galaxies. The clusters, which are giant dark-matter haloes filled with galaxies forming the nodes of the cosmic web, are immersed within an extremely hot gas: essentially a plasma. This plasma formed when primordial intergalactic gas flowed into the dark matter halo, becoming heated to very high temperatures – millions of degrees. The exact physics of this heating is quite difficult to explain, but to put it simply, the energy of the gas in the cluster environment increases according to the total gravitational potential – determined by the total mass – of the cluster. Just as the galaxies within clusters get accelerated to high velocities, so the gas gets whipped up too.

This intracluster gas is so hot that it emits x-ray radiation. Telescopes sensitive to x-rays, like Chandra and XMM-Newton, can detect this radiation, and these telescopes see clusters not as dense swarms of galaxies but as big bright blobs of x-ray emission on the sky. The hot gas filling the cluster has a rather important effect on the galaxies within it. One of the most dramatic effects is called ram-pressure stripping, which is the galactic equivalent of blowing out a candle. Imagine a galaxy like the Milky Way hurtling into a dense

This is an image of one of the most massive clusters of galaxies detected in the early universe (the light travel time from this object to Earth is about 7 billion years: the Earth and the solar system were not yet formed when the light we are seeing now left this cluster). It is called El Gordo, or 'the big one'. The blue haze shows the emission of x-rays, which come from the very hot gas that fills the 'intracluster medium' within clusters such as this. The intracluster medium forms as primordial and intergalactic gas is attracted to the overall gravitational potential of the huge dark matter halo that is present here. As the gas is accelerated towards the potential, like a bowling ball released from the top of a hill, it gets heated to high temperatures – tens of millions of degrees – enough to emit x-rays and too hot to collapse into galaxies. El Gordo is actually two clusters in the process of merging. Just as two galaxies can merge, so even massive structures like clusters can merge and combine, brought together by the force of gravity. One of the key facets of our picture of the growth of galaxies and the structures they inhabit is the idea of 'hierarchical growth', whereby large objects can grow through the coalescence of smaller ones.

The cluster Abell 2744 showing the x-ray emission of hot gas between the galaxies (purple) and the distribution of dark matter (blue) revealed by a gravitational lensing analysis. The immense mass of the cluster (a large fraction of which is dark matter) distorts the images of galaxies behind it, and this distortion can be used to map out the dark matter.

cluster. The cluster accelerates the galaxy to a high velocity, up to hundreds of or even 1,000 kilometres per second. But the disc is not traversing empty space; it is crashing through a dense, hot medium: the cluster atmosphere. This exerts pressure on the galactic disc, buffeting and ramming the gas within it. If the pressure is too much, then the gas that is only loosely bound to the disc gets ripped out, trailing behind the inbound galaxy like a comet's tail. Progressively, as the ram pressure increases closer to the core of the cluster, more and more gas is removed from the hapless galaxy, peeling away like an onion. Without cold gas in the disc, there can be no more star forma-tion, and so at its most extreme, ram-pressure stripping can actually shut down star formation in gas-rich galaxies that enter cluster environments, as they are dragged into the hostile environment through gravity.

Ram-pressure stripping is not the only force at play in the star-formation histories of cluster galaxies: the hot cluster atmosphere also makes it hard for any new, fresh intergalactic gas to collapse onto the galaxy – they are 'starved' of gas, and so eventually cease star formation as the internal reservoirs are consumed without replenishment. A cluster galaxy's destiny is – generally – to become a dead, passive galaxy, reddened with the colour of a geriatric stellar population. So when we look at the galaxies within the centres of clusters, we find what is termed the 'red sequence': galaxies with a range of stellar masses but all with a very similar red colour, indicating a mature, passively evolving stellar population. These galaxies will just sit there for aeons, sloshing about within the cluster potential but otherwise living an uneventful life. Most of these galaxies' 'exciting' (if that's the right word) phase of evolution has now passed.

This image of cluster Abell S0740 neatly shows several galaxy types. The view is dominated by a large, bright elliptical galaxy, which is effectively a giant ball of stars. Next to the elliptical can be seen spiral and lenticular galaxies, and many others in the background. The elliptical galaxy is quite featureless compared to the spirals: it has a very smooth stellar distribution, no dust lanes can be seen and all of the stars are of a uniform hue. This is because ellipticals tend to be 'dead' – no longer forming new stars, with little remaining gas. The uniform red colours of the stars indicate an old stellar population: the majority of this galaxy's evolution was in the past, with mergers and interactions likely to be important in shaping this galaxy. Galaxy mergers can destroy rotationally supported discs, sending stars on random orbits and resulting in a more bulbous 'pressure supported' morphology.

Understanding the evolution of galaxies requires us to look back into the past. The universe did not always contain galaxies in the form that we see them today. The structures we identify as galaxies – distinct entities of dark matter, stars, gas and dust bound by gravity – had to form and evolve over time, emerging as complex structures from a hot, *almost* uniform mixture of matter that existed shortly after the start of the universe. The formation of galaxies required that the hot primordial mixture, which contained the basic elements hydrogen and helium (and smaller amounts of the light elements

deuterium and lithium), cooled and collapsed into dense clumps. If there was no mechanism to do this, then single atoms of hydrogen could not conglomerate in clouds, which could then not bind together to form molecular clouds, which in turn could not form stars via nuclear fusion. In short, galaxies could not form. But galaxies *did* emerge from that early maelstrom. During the journey from that initial formation shortly after the Big Bang to the present day, the properties of galaxies have changed, and tracking these changes is one of the goals of extragalactic astronomy. One of the key changes is in the rate of growth of galaxies, as reflected by their star-formation rates. This is the next step in our story.

This is the heart of a rich cluster of galaxies, dominated by a population of red-yellow elliptical and lenticular galaxies. The streak of light next to the largest elliptical galaxy to the right of this image is a more distant galaxy along the same line of sight, whose light has had to pass through this cluster on its journey to us. During that process, the warping of space-time caused by the large mass of the galaxy cluster (both the stellar matter you can see in this image and the intracluster gas around the galaxies and the 'halo' of dark matter these galaxies reside in) has distorted and magnified the light of that background galaxy: it has been gravitationally lensed. To the top left, we can see a galaxy that seems to be disturbed. Blue splotches and streams emanate away from it like a comet. This is ram-pressure stripping in action. As a galaxy passes through a rich cluster like this, it encounters a hot atmosphere of gas: a plasma. This atmosphere is not visible in this image because it does not emit visible light; it is best seen with x-rays. Nevertheless, the *effects* of the atmosphere on the galaxy can be seen: in the same way that an umbrella can be blown back by a strong wind, so the hot plasma exerts a pressure on the disc of this galaxy, stripping out gas that trails behind. The disturbance can cause dense patches of gas to collapse and form stars, as is evidenced here by the blue colours of the streams. Note that the elliptical galaxies tend to have very little or no gas, so the consequences of ram-pressure are not as notice-able. This is one example of the effect a galaxy's local environment can have on its evolution.

A selection of galaxies viewed in the infrared bands with the Wide-field Infrared Survey Explorer (WISE) satellite, which has mapped the entire sky at wavelengths of 3.4, 4.6, 12 and 22 microns.

FOUR

The Evolution of Galaxies

Hale Pohaku is the residence for visiting astronomers using any one of the many telescopes perched atop Mauna Kea, 4 kilometres above the waves of the Pacific Ocean. Mauna Kea is the mountain that, along with its neighbour Mauna Loa, dominates the Big Island of Hawai'i. Hale Pohaku, or HP for short, is at an altitude of 2,700 metres, some way below the summit. Here it is comfortable enough to eat, sleep and work when not manning the telescopes, and below the 'danger zone' for acute altitude sickness. From sea level you can sometimes see the white domes glinting in the sunlight on the summit, but often the foothills are enshrouded by thick cloud, brought on as warm, moist, Pacific-laden air is driven up the mountain. At the foot of Mauna Kea the air is positively thick, but travel up the mountain and the air thins, the sky clears and you are 4 kilometres closer to the stars.

One of the telescopes I use on Mauna Kea is the James Clerk Maxwell Telescope (JCMT), and in particular the second-generation Sub-millimeter Common User Bolometer Array (SCUBA). The first SCUBA became defunct a few years ago, and SCUBA-2 represents a large technical improvement in this area of astronomy. It is designed to detect light at the submillimetre wavelengths of about 0.45 mm and 0.85 mm. These particular wavelengths are not arbitrarily chosen; it's the Earth's atmosphere that dictates what submillimetre wavelengths we can see, because for the most part the atmosphere is very effective at absorbing infrared and submillimetre-wave light. There are narrow 'windows' in the transmission of the atmosphere where photons of certain frequencies *can* get through, however, and two of these windows are at these special wavelengths that SCUBA-2 sees. Still, the best transmission is achieved when there is very little water vapour in the air, and these conditions are best found in places like Mauna Kea and the Chilean Atacama.

Extreme galaxies, hidden by dust

Submillimetre light is radiation between the infrared and radio bands in the electromagnetic spectrum. JCMT is unlike an optical telescope, but similar to a classic radio antenna: its 15-metre-wide collecting dish is not made of polished glass but 276 panels of aluminium, which is perfectly capable of capturing the submillimetre photons. These are reflected to a smaller, secondary mirror and then to a detector (in this case SCUBA-2, but JCMT also has other instruments). As with many other telescopes, JCMT is housed by a dome that protects the telescope itself, the supporting infrastructure (computers, electronics, cryogenic equipment and so on) and a control room. The dome shields everything from the outside world, which is particularly important at the summit of Mauna Kea, where environmental conditions can get seriously harsh, with sub-zero temperatures and gale-force winds screaming across the mountain. The dome has an opening that allows the dish to see the sky, and the whole thing can rotate around in order to point to different parts of the celestial sphere. When the dome opens, there remains one last physical barrier between the dish and the sky: the world's largest piece of GoreTex. The GoreTex is a further protective layer, and happens to be 97 per cent transparent to submillimetre photons. If our eyes were sensitive to submillimetre photons then this grey, opaque sheet would look like a pane of glass.

Right now, SCUBA-2 is a relatively new instrument, allowing us to perform some pretty incredible science that was simply not possible a few years ago. The new power comes from the sensitivity of the detectors, allowing us to more efficiently record submillimetre photons, and also its size: the camera is much larger than its predecessor, and this makes it much easier to make large maps of the sky; this is essential for survey work. Unfortunately, the astronomical photons that we're trying to detect – from distant galaxies – are swallowed in a deep, turbulent sea of ambient radiation. Emission from the atmosphere and even the thermal emission of the telescope itself is the dominant signal that the camera sees. Any astronomical signal is a tiny blip among this background, so these contaminating components must be subtracted before we can make a scientific image. Luckily we have clever software to do this, and those dominant signals can be very effectively modelled and removed by filtering the data each bolometer records as SCUBA-2 scans across the sky. But *why* are we observing these submillimetre photons, and not some other wavelength?

We have seen how galaxies can be actively forming new stars, and that this activity is seen at ultraviolet and visible wavelengths – direct emission from the new, massive stars themselves, and also through the emission lines emitted by ionized gas around those stars. The flux we measure in, say, the ultraviolet band, can be converted to the galaxy's star-formation rate because we know the number of ultraviolet photons being emitted by those young stars. Similarly, the total number of, for example, H-alpha photons being emitted by the ionized gas in the concomitant HII regions is also directly related to the number of ultraviolet photons being emitted by the newly formed stars. But we've also seen how interstellar dust can block out this light, rendering our estimates of the star-formation rate inaccurate. We call this effect 'extinction' because of the deleterious effect on measured flux. Galaxies that are vigorously forming stars tend also to be very dusty, or at least their star-forming regions are, and the light emerging from the new stars is therefore largely absorbed and scattered by the dust. This is a problem if we really want to understand how the star-formation rates of galaxies vary between galaxy types and across cosmic time.

What is the 'dust', exactly? We're talking about sub-micron-scale particles – grains – mainly of carbon and silicon, about the size of the motes that make up cigar smoke, but far more rarefied. This material is produced naturally during the late stage of stellar evolution in the atmospheres and circumstellar environments of stars, and gradually gets spread throughout interstellar space when the stars die, either as they eject their atmospheric layers in a nova or via a more violent dispersal in a supernova. For this reason, dust is usually clumped in dense patches nearby or within the sites of new star formation – the very regions we want to study if we are to measure the star-formation rate. There is a solution to this problem, however, and that is to turn to parts of the electromagnetic spectrum that can actually detect this dust directly as it heats up through the absorption of the light of the nascent stars.

When a dust grain absorbs an ultraviolet photon, it heats up, gaining thermal energy in the form of oscillations of the atoms within. Like microscopic glowing coals, dust grains being irradiated by ultraviolet photons suddenly become visible in the infrared part of the electromagnetic spectrum as they re-radiate the thermal energy. The amount of re-radiated energy is proportional to the amount of incident radiation that came from the young stars, solving the problem of how to measure the star-formation rate of a galaxy that is heavily blanketed by dust. This was a cue for astronomers to

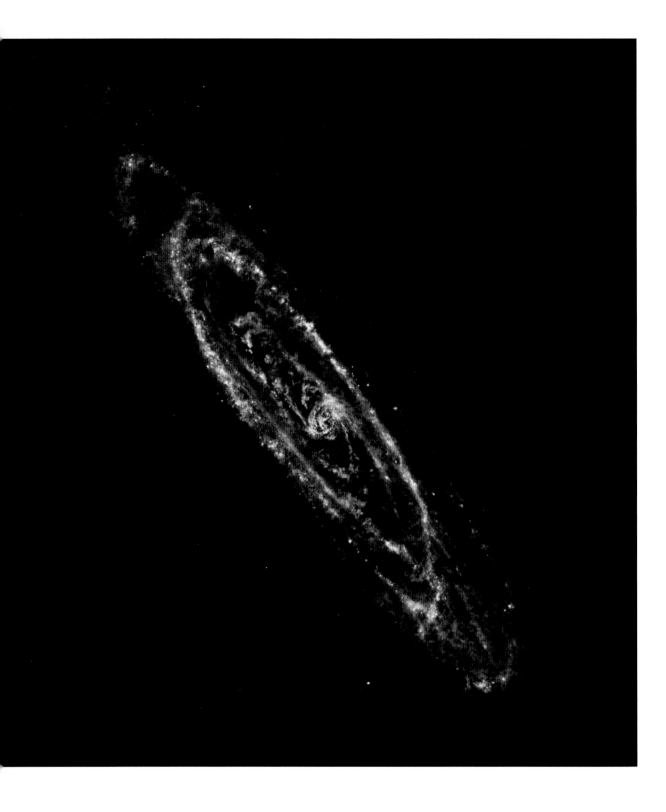

The Wide-field Infrared Survey Explorer (WISE).

An image of the Andromeda galaxy at far-infrared wavelengths taken by the Herschel Space Observatory. Far-infrared emission traces cool dust, which in galaxies like M31 is concentrated in the spiral arms, clearly illustrated in this example. Spots where the emission is brighter indicate the location of dense star-forming regions, where dust blanketing the gas clouds in which new stars are being formed glows brightly.

develop telescopes and other instruments to detect and map this infrared emission.

The spectrum of radiation emitted by the majority of interstellar dust is close to what in physics parlance is called a 'black body'. A black body is an object that absorbs all of the electromagnetic energy (photons) that hit it, and, if it remains at a constant temperature (we call it in 'equilibrium'), re-emits the radiation across a range of frequencies with a very characteristic spectrum called a Planck function (after quantum mechanics pioneer Max Planck). The black-body spectrum has a peak at a specific frequency, or wavelength, of light that corresponds to the temperature of the body (the same principle we have encountered when looking at the colours of stars, which can also be described as black bodies). Now, the typical temperature of interstellar dust is fairly cold by our standards, a few tens of degrees above absolute zero. This temperature corresponds to radiation in the 'far' infrared part of the spectrum, around 100 to 200 microns.

Unfortunately, most of the infrared light emitted by astronomical sources is blocked out by the Earth's atmosphere, save for the narrow 'windows' we talked about above, such as those in the submillimetre bands that SCUBA-2 operates in. Therefore, infrared observations are best conducted from space. The most successful infrared telescope of recent years is the Spitzer Space Telescope (named after astronomer Lyman Spitzer), one of NASA's Great Observatories. Spitzer carried (or rather, still carries) instruments that could detect radiation at around 4 to 160 microns, and carried out some outstanding science. A good example is the Spitzer Infrared Nearby Galaxies Survey (or SINGS) project – where, as the name suggests, Spitzer was tasked with imaging local galaxies in the infrared to map and understand the distribution of their dusty interstellar media and the nature of their star formation, and generally learn more about the infrared emission of local, well-studied galaxies in unprecedented detail. When we look at images of spiral galaxies taken by SINGS, and compare the infrared emission with the optical light, it is totally clear how the dust – which appears as dark occluding patches in the optical – becomes visible in the infrared, tracing the obscured star-forming regions.

All infrared detectors must be kept cool with cryogens, and since it is a space telescope (not actually in orbit around the Earth, but in an Earth-trailing orbit around the Sun), Spitzer's cryogen supply cannot be topped up. As the cryogen is used up, the instruments that need to be cooled shut down. The only remaining working instruments on the satellite at the time of writing are the two shortest wavelength cameras of the InfraRed Array Camera (IRAC), working at 3.6 and 4.5 microns in wavelength. Fairly soon, they too will become defunct, and Spitzer will have completed its mission. Between 2009 and 2013, the Herschel Space Observatory, which operated at longer wavelengths of about 50 to 500 microns, completed its mission when its cryogenic helium supply ran out. One of Herschel's great achievements was to perform very large area-mapping surveys, detecting thousands of galaxies bright at far-infrared and submillimetre wavelengths, as well as detailed studies of local galaxies in the far-infrared that complement the work done by Spitzer.

Re-radiated infrared light is such an important source of galaxy emission that, when averaged over the lifetime of the universe, about one out of every two photons generated from star formation is emitted in the infrared. It's a big part of the extragalactic energy budget.

What it means is that about half of the star-formation activity in the universe is actually traced by infrared-emitting dust, rather than direct ultraviolet and optical emission from stars and gas. That's averaged over all galaxies. When we consider certain individual galaxies, we find that the most extreme systems in the universe – the most actively star-forming ones, for example – are often *dominated* by their infrared output,

The spiral galaxy M100 viewed in the mid-infrared part of the electromagnetic spectrum, at wavelengths of 3–8 microns, from the Spitzer Space Telescope. The mid-infrared bands trace warm dust associated with star-forming regions. Unlike in optical images, where dust blocks out the light from stars, at infrared wavelengths the dust itself glows (red in this image) while stars are much fainter (appearing here as the faint blue haze). The central region of this galaxy is glowing brightly where there is a ring of rather intense star formation, and the dense spiral arms are clearly highlighted with many bright 'knots' of activity. Infrared observations provide a complementary view of galaxies, which is essential given the ubiquity of interstellar dust.

Another mid-infrared view, this time of the famous Sombrero galaxy, again imaged by the Spitzer Space Telescope in the mid-infrared bands (3–8 microns). The Sombrero is slightly tilted from edge-on from our vantage point, and its most striking characteristic is a large, dusty disc, appearing as a ring surrounding a rather elliptical distribution of relatively old stars. In this infrared view, that ring of dust glows red as it is warmed by starlight.

obscured so much that they can be almost invisible in the visible light bands. The submillimetre bands are critical because they allow us to measure part of the far-infrared spectrum of galaxies at wavelengths reasonably close to the spectral peak, giving us a good measure of the total infrared luminosity, and therefore the star-formation rate. Sometimes the study of galaxies at infrared and submillimetre wavelengths is called the exploration of the 'dusty' universe.

This exploration really began in earnest back in 1983, when a space-borne observatory called the Infrared Astronomical Satellite (IRAS) was launched.

A galaxy known as the Sunflower (M63), imaged by Spitzer in the mid-infrared, showing warm dust tracing the intricate spiral arms. Due to the relatively long wavelength of light that Spitzer sees (compared with visible light), combined with the small telescope aperture (less than 1 metre), these images are not as sharp as Hubble's – the resolution is poorer. Nevertheless, Spitzer has provided some of the most important data on many aspects of galaxy properties, formation and evolution over the past decade.

Spitzer's mid-infrared view of the galaxy Arp 77, a spiral galaxy with a prominent bar structure passing through the core and joining the spiral arms. Notice the bright ring of infrared emission at the core of the galaxy; this is the hot dust emission associated with a ring of star formation surrounding the active central black hole. The bar structure is in part responsible for this nuclear activity – it can serve to transport gas and stars from the disc towards the core of the galaxy.

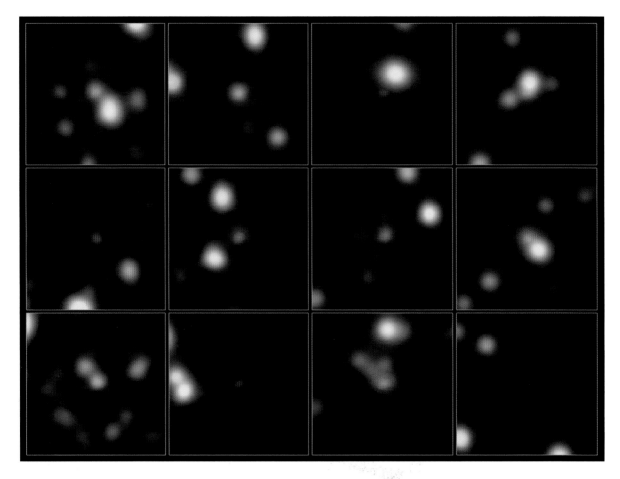

The red/orange blobs at the centre of each panel are very distant galaxies that emit most of their light in the submillimetre part of the electromagnetic spectrum. One of the main achievements of the field of galaxy evolution is the discovery that the overall rate of galaxy growth, traced by the average star-formation rate, was much higher in the past. Some of the earliest galaxies were forming stars at remarkably high rates, hundreds of times faster than the rate at which the Milky Way forms new stars. Galaxies undergoing intense 'bursts' of star formation also contain large amounts of obscuring dust that blocks out most of the visible light from stars. Instead the dust absorbs the starlight and heats up in the stellar radiation field. This energy is 're-radiated' in the far-infrared part of the electromagnetic spectrum, and due to the effect of redshift – the increase of the wavelength of light observed from distant galaxies due to the expansion of the universe – is readily detected in the 'submillimetre bands'. Galaxies in this category are called submillimetre galaxies, or SMGS. Why is this class of galaxy interesting? One theory is that SMGS are the progenitors of massive elliptical galaxies today, but seen at a time when they were forming most of their stellar mass, so studying distant SMGS could provide clues as to the physics of formation of the most massive galaxies in the cosmos.

It was the first telescope to map the entire sky at infrared wavelengths of light (more precisely, at wavelengths of 12, 25, 60 and 100 microns), and during its ten months of scientific operations, IRAS opened up a brand new window onto the universe by giving us access to a portion of the electromagnetic spectrum that was poorly explored until then. IRAS was paving a rich new avenue of research into galaxy evolution that three decades later is still accelerating, and actually forms a significant part of my own research.

Perhaps the most important achievement IRAS made was the discovery of thousands of galaxies that had not been catalogued before. This was possible because, although they are relatively faint at optical wavelengths and were therefore missed or overlooked by previous surveys, they glow brightly at infrared wavelengths. IRAS also mapped the large-scale distribution of interstellar dust in our own galaxy: another of the satellite's key discoveries was that, in almost every direction it looked, there was always some amount of wispy, diffuse infrared emission on the sky. This was termed 'cirrus', like the high clouds we see on Earth, but in this case, instead of water vapour, IRAS was detecting the pollution from our galaxy's star-formation history: the dusty detritus formed in previous generations of stars.

This galactic cirrus poses a problem for extragalactic studies at ultraviolet and visible frequencies, because before contending with the gauntlet of absorption presented by the Earth's atmosphere (where water and other molecules can easily snuff out photons that are passing though), the ultraviolet and visible-light photons from distant extragalactic sources must pass through the rest of our own galaxy. If a photon coming from some distant galaxy encounters some of the dust within our *own* galaxy, then it can also get absorbed. This is called galactic extinction, and we have to correct for its dimming and reddening effect on the observed intensity of extragalactic light.

To make the extinction correction, we need detailed maps of where the galactic dust is and how thick it is (we can get this from the all-sky maps made at infrared wavelengths, for example). Combined with a formula for expressing how strong the absorption is for different frequencies of light (called a 'reddening law'), we can add back the emission that the galactic dust took away. In some directions, like the galactic plane, the extinction is so extreme that no extragalactic light can make it through. So, when we perform very deep observations of the distant universe, ideally we want to peer out of the disc of the galaxy where the amount of intervening cirrus is low, so the galactic extinction of extragalactic light is minimized. This is another disadvantage of being an extragalactic astronomer living within the disc of a galaxy. It's curious to think, however, that astronomers in other societies in the galaxy will be able to explore the extragalactic universe to varying extents, depending on whether they reside in a very dense part (such as closer to the bulge), or in the galactic suburbs – perhaps even in one of the Magellanic Clouds.

The other main achievement of IRAS was the discovery of a population of galaxies that were emitting huge amounts of infrared radiation: in excess of 1,000 billion times the luminosity of the Sun. These were termed ultra-luminous infrared galaxies (ULIRGs). Along with quasars, these galaxies rank among the most luminous objects in the universe. Although some of the galaxies were previously known from optical surveys, they weren't considered to be particularly special. Only when IRAS revealed their intense infrared output did people start taking notice.

What is powering the very intense activity in these particular galaxies – why are they so much more extreme than, say, our own Milky Way or nearby M31?

The violence of galaxies

On closer inspection, virtually every ULIRG in the local universe was found to be either irregular or disturbed in shape, or in the process of a 'merger': the gravitational coalescence of two galaxies. Galaxies are not rigid structures; they are more like drops of fluid that can be squished and squashed by gravitational forces. During the process of galaxy–galaxy collision, strong 'tidal' forces, exerted on the gas and stars by the mutual gravitational attraction between two (or more) galaxies, dramatically distort the shape of each galaxy. For two spiral galaxies, for example, this usually involves stars and gas in the spiral arms being ripped out into long trailing filaments as the galaxies first approach and then fly by each other, a process that might be repeated several times depending on the relative velocities involved: a dance, if you like, orchestrated by gravity and culminating in coales - cence. Sometimes, instead of overshooting each other with a glancing blow, pulling along stars and gas and dust in the altercation, two galaxies can smash right through each other, dramatically redistributing matter in the galaxies, sometimes in spectacular fashion, such

Detail of one of the arms of the so-called Meathook galaxy, named after the appearance of its distorted spiral structure, thought to be caused by a previous gravitational disturbance due to the passage of another galaxy. Like the foam on the crest of a wave, the arm is effervescent with lots of young, blue stars.

as in the formation of 'ring' galaxies. Eventually both galactic systems 'sink' to the common potential well that represents the merged dark-matter haloes in which the galaxies reside.

You can make a simple model of the dynamics of galaxy mergers in your kitchen with a couple of marbles, or any kind of rolling sphere, and a large bowl. Set those two marbles rolling down the side of the bowl and they will follow a similar pattern to two merging galaxies, their motions governed by their starting velocities (give one marble an extra boost down the side of the bowl and see the effect), and the shape and strength of the gravitational potential, here modelled by the depth and shape of the sides of the bowl.

During mergers, gas in the discs gets disturbed and compressed, and this can promote the collapse of giant molecular clouds as shock waves and turbulence propagate through the system, introducing density perturbations in the gas. This is the perfect storm for the triggering of new star formation, as these density perturbations rapidly grow under the *local* forces of gravity (gravity works on all scales, driving the motions of the galaxies that are merging, and the internal behaviour of the gas and stars as well) and eventually ignite in star formation. Towards the finale of a merger, a large fraction of the disc gas can get driven to the nuclear regions of the merged galaxy, forming a dense molecular complex spanning perhaps a few thousand parsecs. When driven to high densities, and when a vast reservoir is available, molecular gas can fuel star formation at a very high rate – up to hundreds to thousands of solar masses of new stars per year. We call this a starburst.

Within a few million years, the many massive young stars that have formed start dying off in supernovae, and begin to pollute the galaxy with lots of dust (in addition to the interstellar dust that was already formed in each galaxy). Thus intensely star-forming galaxies are often heavily obscured, with a large fraction of the dust co-located with the active sites of star formation, since massive stars die close to the stellar birthing grounds, not having enough time to migrate away. So, galaxies like ULIRGs are assembling stellar mass in furious episodes of star formation, but most of this activity is hidden behind a dusty cloak. The cloak heats up and emits far-infrared radiation, and this is how IRAS detected them.

A classic – and my favourite – example of an ongoing merger is the aptly named Antennae Galaxies. In our 1-metre box model of the local universe, the Antennae Galaxies would be about 70 centimetres from the Milky Way, equivalent to about 14 megaparsecs. You can still make out that these used to be two galaxies that probably looked like fairly typical spirals, but they have collided, grossly distorting their once regular morphologies. The two buckled discs are now co-mingled, forming two dense clumps of stars, gas and dust, peppered with the distinctive glow of HII regions forming new stars, likely ignited by the tidal forces exerted on the gas clouds. Stretching away from the coalescing core are two long stellar streams – the antennae – ripped out of the discs of the galaxies during an earlier stage of the merger when the galaxies passed close by each other. Entwined in a gravitational dance, the two galaxies are becoming one; a process that probably lasts about a billion years. With the Antennae, and other galaxies at different stages in the merger dance, we get to see this astrophysical process 'in the act'.

A collision between two bodies of similar mass – like the future collision between M31 and the Milky Way, for example, or the ongoing one between the Antennae Galaxies – is called a 'major' merger. Major mergers induce dramatic evolutionary changes, obviously increasing galaxy mass, triggering new star formation and black-hole growth (since driving gas to the central region will naturally provide the fuel from which a central black-hole can grow), transforming morphology and enriching and mixing the interstellar medium, spreading the heavy elements around. On the other hand, 'minor' mergers of a big galaxy and much smaller systems are more frequent; partly this is because lower-mass galaxies are far more numerous than high-mass systems (the distribution of galaxies of different mass is called the mass or luminosity function), and also because lower-mass systems often hang out around larger galaxies as satellites, increasing the chance of coalescence. Again, we can look to our own galaxy as an excellent example. We have talked about the two largest galactic satellites, the Large and Small Magellanic Clouds, but the Milky Way is also surrounded by a posse of at least a few tens of lower-mass dwarf galaxies at distances of between ten and a few hundred thousand parsecs. These dwarfs are given names according to the constellation they can be seen in (remember, this is just a projection effect; the satellites are far more distant than the stars in those constellations), so we have the likes of the Sagittarius Dwarf, the Draco Dwarf and Leo IV. New dwarfs are being discovered every few years; despite their proximity, they are actually quite challenging to spot because of their low mass (and therefore low surface brightness) and the fact that they are spread out over large areas of the sky, thus requiring large area surveys if we are to find them.

Two galaxies engaged in a dance: Arp 273 is a pair of interacting spiral galaxies brought together by gravity. Gravitational interaction is exerting tidal forces on the stars and gas and dust, distorting the structure of each galaxy. Eventually the pair will coalesce into a new galaxy, redistributing the stars and gas in the process. The interaction can cause the collapse of clouds of gas in the discs of the galaxies, and so galaxy interactions and mergers will often 'trigger' new star formation.

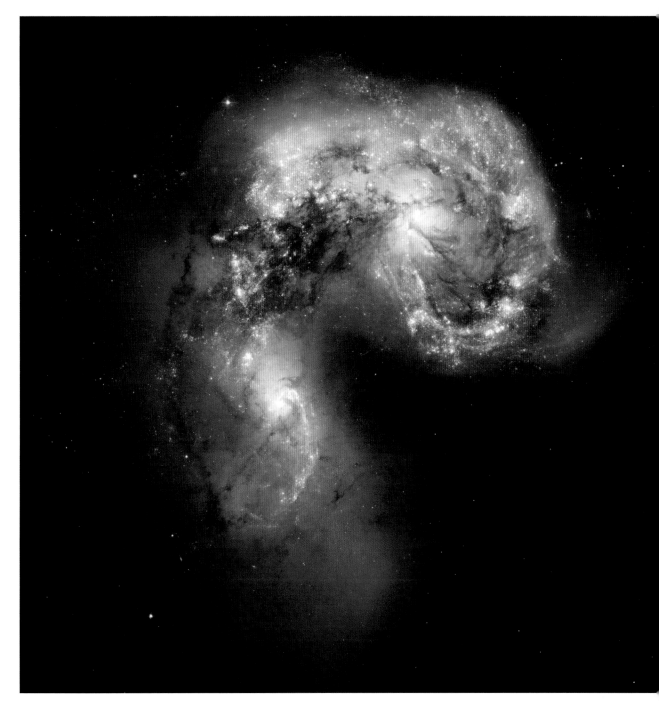

Another famous interacting pair of galaxies: the Antennae. This view concentrates on the nuclei of the two galaxies, which are close to coalescence. Red light shows the presence of ionized hydrogen, associated with new stars that are bursting into life as the gravitational interaction stirs up and compresses cold gas that was originally in the discs of the two galaxies. Many newly formed stars can be seen in blue light and, as always, interstellar dust obscures some of the optical light, especially in the regions where star formation is most extreme.

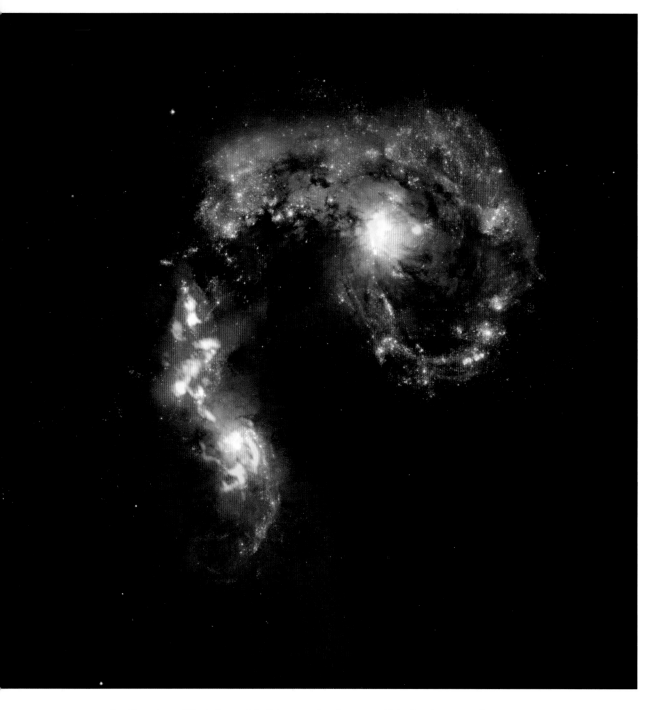

Another view of the twin nuclei of the merging Antennae Galaxies, incorporating data from the Atacama Large Millimetre Array, which operates at submillimetre and millimetre wavelengths of light. The splotchy red colour corresponds to dense molecular gas, overlaid on a visible light image. Notice that the molecular gas is located in regions where there are large amounts of interstellar dust, which is blocking out starlight. Dust and gas tend to be co-mingled in the densest parts of the interstellar medium of galaxies; in this case those gas clouds are being tugged and compressed by gravitational forces during the violent galaxy merger, which can trigger new bursts of star formation.

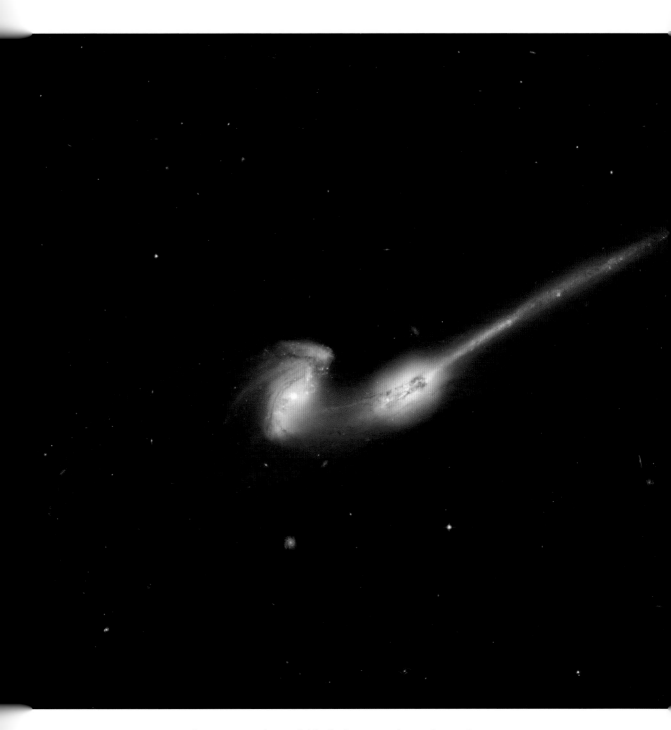

A pair of interacting galaxies dubbed 'The Mice', destined to coalesce into a single, merged galaxy in the future. Most striking are the long, blue, tidal tails of stars, ripped from the discs of the galaxies by gravitational tidal forces during the interaction. The two blobs of yellow/orange stellar emission represent the central bulges of the two galaxies, which retain more of their shape during the merger; nevertheless, the stellar envelopes are starting to overlap and amalgamate.

Arp 116 is a pair of interacting galaxies in the Virgo cluster of galaxies, comprising an elliptical galaxy and a spiral galaxy. This gravitational interaction is thought to have just begun, but the image nicely contrasts the morphological characteristics of galaxies: elliptical galaxies are giant, spheroidal congregations of old stars, generally not forming any new stars. Spiral galaxies are arranged in a disc (in this case the spiral galaxy is nearly face-on) and tend to be actively forming new stars.

The interacting pair of galaxies Arp 87, bridged by a delicate tether of stellar light and destined eventually to entwine and merge into a single galaxy. Evidence of recent star formation triggered by the interaction can be seen in the blue starlight wrapped around the galaxy on the left and in the disturbed spiral arm of the galaxy on the right, which is encrusted with bright, jewel-like points of light: clusters of new stars.

A tidally disrupted galaxy which for obvious reasons is known as 'The Tadpole'. In this case, a smaller galaxy ploughed into a larger spiral, leaving a wake of stars in a long, trailing stream, teased out through tidal forces. Within that stream can be seen several clusters of massive blue stars, indicating patches of active star formation in the tail (which also contains cold gas, not visible at these wavelengths). Some of these star clusters might be destined to become globular clusters that will eventually be attracted to, and surround, the main body of the galaxy. The background shows thousands of other, more distant galaxies within the field of view.

As they orbit our galaxy, the dwarfs can get ripped apart and entrained into long tidal streams as they pass close to the main bulk of the galaxy. The Sagittarius Dwarf actually appears to be associated with a long ribbon of stars and gas that probably encircles the disc of the galaxy; the lost material once in the satellite that has been drawn out like a piece of gum during a passage around and through the main disc. The exact motions of satellites around the Milky Way is (thought to be) governed largely by the structure of the dark-matter halo the disc is contained within. Therefore, by studying the pattern of things like stellar streams that are associated with dwarfs and betray the past motions of the satellites, we can learn something about the distribution of mass of the dark structure within which the luminous component of the galaxy is embedded.

The Milky Way is obviously fairly well evolved, and the accretion of the occasional satellite is not going to radically change things; the dwarfs' masses are small compared with the mass of the Milky Way itself. However, the idea that massive systems can grow from the blending of smaller ones is a key facet of our current model of how galaxies form and evolve in the universe, especially in the early universe. This is called the hierarchical paradigm for two reasons; first, because we can think of galaxies belonging to a 'hierarchy' of structure on many spatial scales. For example, a disc galaxy surrounded by dwarf satellites might belong to a small group of galaxies which itself forms part of a filament of galaxies and groups that link to a massive cluster of thousands of galaxies. Second, one of the simple models of galaxy growth in our current model is that bigger galaxies can be assembled from the progressive merging of smaller systems in a process called 'bottom-up' formation. In fact, the exact details of early galaxy formation and growth appear to be more complicated than this; for one thing, observational evidence seems to suggest that the most massive galaxies seen today formed most of their stars earlier than lower-mass galaxies. In the naive interpretation of the bottom-up scenario, one might think that the most massive galaxies are the last things to form, since first you need to assemble the smaller building blocks. It turns out that the exact details of galaxy formation in the early universe are more complicated than a simple bottom-up progression, and one of the main areas of research currently is *exactly how* the gas and dark matter form and flow into structures – dark-matter haloes. The mixture of physics governing this process is complex.

Nevertheless it's very clear that throughout the history of the universe, mergers have been important in shaping the galaxy population, since a significant fraction of galaxies will go through a merger event at some point in their lives. Whereas major mergers make a complete mess of the spiral disc, and completely transform galaxies in a very obvious way, the impact of minor mergers such as the interaction with in-falling satellites is more subtle – barely noticeable, in fact, save for perhaps a slight warp in the disc caused by the gravitational perturbation, or a small burst of star formation as fresh gas is accreted onto the disc. As the Sagittarius Stream bears witness, however, the effects on the dwarf itself are devastating – often it is ripped apart completely, cannibalized by the host. While a typical image of a distant spiral galaxy may at first glance look 'normal' and undisturbed, extremely deep observations that collect a lot of light, probing to very low surface brightness, often reveal the faint glow from diffuse stellar streams that are surrounding and entwining the galaxy like a cage of light. These are echoes of the ongoing saga of galaxy evolution, illustrating again how galaxies are not static and constant, but complex and dynamic environments.

The name given to many ULIRGs, like the ones discovered by IRAS, is 'starburst galaxies': galaxies forming stars at rates of up to a several hundreds of times higher than our galaxy (although it turns out that some ULIRGs' infrared output comes from an active galactic nucleus – a growing super-massive black hole – rather than star formation). Starbursts are an important species in the taxonomy of galaxies. One of the most famous local starburst galaxies is M82, also known as the Cigar Galaxy because of its shape. M82 is about 3.5 megaparsecs away, about 18 centimetres from the Milky Way in our 1 metre box model, and part of a small group of galaxies that includes the spiral galaxy M81 and the smaller galaxies NGC 3077 and NGC 2976 (NGC stands for 'New General Catalogue', or more properly the New General Catalogue of Nebulae and Clusters of Stars, a catalogue of deep-sky objects compiled by astronomer John Louis Emil Dreyer in the late nineteenth century). M82 is an archetypal starburst galaxy, bright in the infrared, but also visible in optical bands, and producing stars at a rate of a few tens of solar masses per year, ten times higher than the Milky Way. Because of its prox-imity, and therefore ease of study, M82 is often used as a prototype and useful comparison specimen for the study of similar galaxies in the distant universe. One of the most striking things about M82 is the 'superwind' emanating from the main body of the galaxy out into extragalactic space.

An image of M82 in optical light reveals a fairly irregular-looking but *almost* disc-like galaxy, inclined to our line of sight. The hazy orange-blue hues

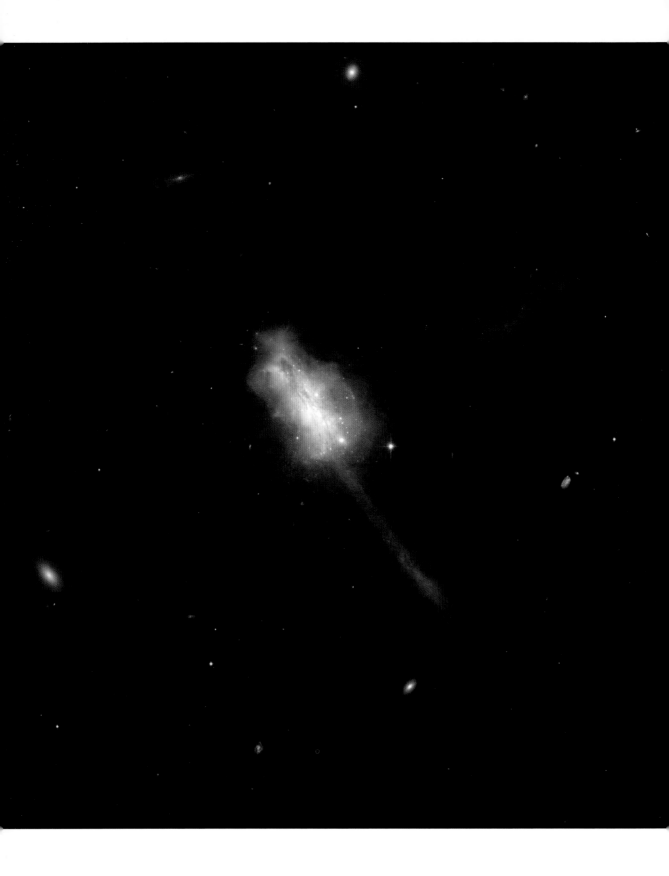

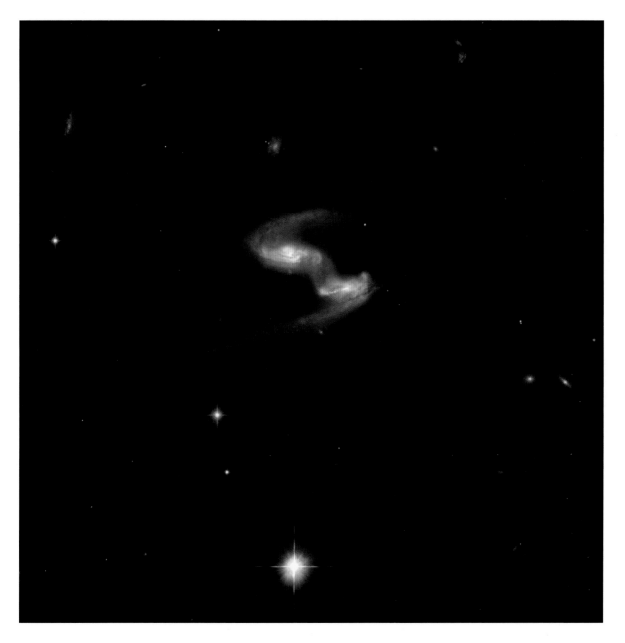

Another merging system, torn apart in a gravitational ballet: interacting galaxy
ESO 77-14.

Like an explosion, the interacting galaxy IC 883 shows the aftermath of a galactic
collision, with stellar shells, wisps and tidal features apparently bursting from a bright,
chaotic mass of stars and gas and dust. This collision has triggered a burst of star
formation, as is common in galaxy mergers, which are important events in the life
histories of galaxies. Many galaxies undergo interaction and mergers, and some
of the most intensely star-forming galaxies in the universe are merger systems.

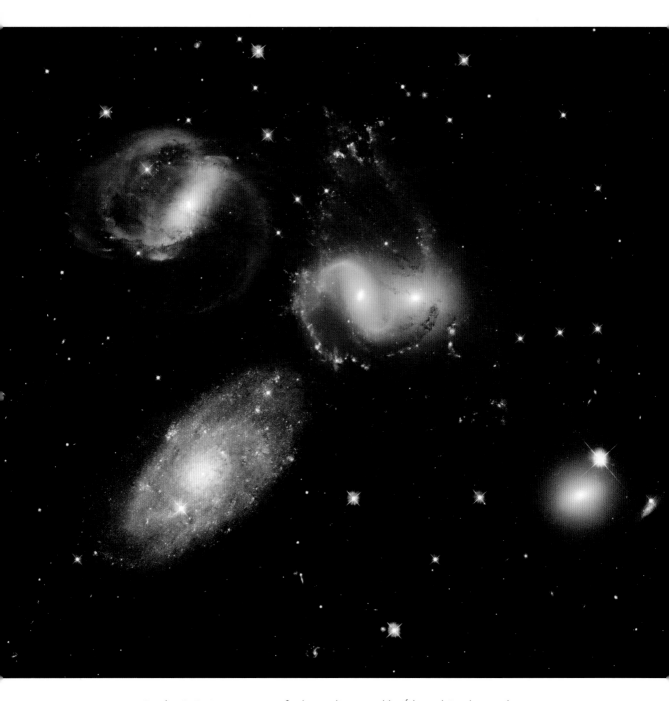

Stephan's Quintet, a group of galaxies discovered by Édouard Stephan in the
nineteenth century. The bluer galaxy on the lower left is not physically associated
with the group, but the three redder galaxies are; they are in the process of
interaction. This interaction is morphologically disturbing the galaxies, tearing
out gas and stars and triggering new star formation in the disturbed material.
This is visible in the fan of stars to the top left of the central red galaxy: patches
of red and blue show the ionized gas (HII regions) and young blue stars forming
in clouds of disturbed, collapsing gas extruded from the disc of the galaxy.

In the early 1980s, Paul Hickson, a Canadian astronomer, compiled a catalogue of 'compact groups' of galaxies. There is a hierarchy of structure in the universe and groups are intermediate-density environments between the low-density environment of isolated 'field' galaxies and the bustling swarms of galaxies inhabiting rich clusters. This is Hickson Compact Group 90, and it is clear that between two elliptical galaxies, another galaxy, probably (what was) a spiral, is being torn apart by gravitational tidal forces, an example of the sometimes violent evolution that galaxies experience in group environments such as this (see Stephan's Quintet).

The irregular starburst galaxy Messier 82, often described as an archetypal starburst galaxy. This image shows the blue/yellow disc where an intense episode
of star formation is occurring within dense gas reservoirs. The red filamentary
emission extending above and below the disc is the light of ionized hydrogen
gas. What we're seeing is an emergent 'superwind': gas and dust being literally
blown out of the galaxy by the energy released – from stellar winds, radiation
pressure and supernovae – at the sites of star formation deep in the galaxy.
Many, if not all, rapidly star-forming galaxies emit these winds at some level,
and they have a role in regulating the rate of galaxy growth by controlling the
amount of gas that can turn into stars. Some of this ejected material will later
'rain down' on the galaxy through gravity, redistributing metals, but in extreme
cases gas can be blown out of the galaxy never to return, like a rocket leaving
Earth's orbit. Understanding the astrophysics of galactic winds and outflows is
currently a key area of research in extragalactic astronomy.

of the stellar population are laced with darker furrows of dust, but actually most of the real action is happening close to the core of the galaxy. However, if we take an image in H-alpha light (so revealing the ionized hydrogen gas) and x-ray light (tracing very hot plasma), what we see is an eruption of hot gas streaming out from either side of the disc in two huge, cone-like prominences. This is called an emergent superwind: gas is being blasted right out of the galaxy and into the circumgalactic medium. Observations at other wavelengths reveal that these streams also contain cooler gas and dust mixed in with the hotter matter, which must have been swept up with the wind as it flowed out of M82 at a million miles per hour.

It is the intense star formation occurring in a compact gas reservoir, deep in the galaxy, that drives this wind. The high star-formation rate results in an enhanced rate of supernovae detonations. When a supernova detonates, a large amount of the energy liberated is kinetic. This exerts pressure on the surrounding medium, sweeping up and driving material (gas, dust and whatever is in the way) like a wind. With many supernovae working in concert, there is a huge amount of energy injection into the interstellar medium. If you combine this with stellar winds driven off individual stars, the result is a gale, blowing through the galaxy. If that wind collides with other material, like gas and dust, it can cause a shock wave, making temperatures soar (evidenced by the x-ray emission and ionized gases). Winds that are strong enough can break out of the galaxy, sending interstellar material into intergalactic space, as is the case with M82.

Galaxies like M82 are rather rare objects today. In fact, when you add it all up, the average amount of star formation occurring in all galaxies in the present-day universe is rather sedate. One of the major findings in the field of galaxy evolution over the past two or three decades has been the discovery that the average rate of star formation in the universe was much *higher* in the past. All astronomical surveys that have measured the average rate of star formation point to the same thing – a steady rise in the rate of production of stars as we look back in time – implying that the rate of galaxy growth has been in decline for some time. To measure the evolution, all you need to do is look at large samples of galaxies seen at different redshifts – remember, the light takes so long to reach us that, when we look at galaxies at high redshift, we're seeing the universe as it was in the past. Galaxies similar to and in fact even more luminous than M82 were more prevalent in the early universe, and these are the sorts of systems we are trying to observe with

SCUBA-2, because they are bright at far-infrared wavelengths, which are detectable in the submillimetre bands.

The star formation history

About 8 to 10 billion years ago, the average rate of star formation in galaxies was about ten times higher than it is today. This appears to be an approximate peak in cosmic activity. If we look to even more distant, and therefore younger, galaxies, it appears that the average star-formation rate gently declines again as we get closer and closer to the Big Bang, towards the time when galaxies first took shape. This is somewhat expected, because galaxies were not always around – they had to form at some point, and presumably 'ramp up' in activity. The experimental constraints on cosmic history (at least, as it pertains to galaxy evolution) before about 10 to 12 billion years ago are much poorer because the uncertainties are large, reflecting the difficulty in performing observations at these very large cosmic distances. We do know that over the majority of the history of the universe, the rate of growth of galaxies has been in sharp decline. This is perhaps the clearest and most important evidence we have of the changes that have occurred in the galaxy population over the course of cosmic history, culminating in what we see around us today (and of course this evolution will continue into the future). The physics governing this galaxy evolution is what people like me are trying to understand.

There are many factors at play in what is responsible for the decline in the global star-formation rate, but one of the main causes is simply the consumption of the gas reservoirs, coupled with a dwindling of fresh supply in galaxies over time. The star-formation rate of a galaxy is closely correlated with the total mass and density of gas within it: more gas means a higher star-formation rate. We know this from detailed studies of local galaxies. In the past, galaxies had more gas because the original reservoirs that condensed when a galaxy first collapsed had not all yet been turned into stars, and the accretion of new gas – from intergalactic space – was occurring at a higher rate. Over time, this supply dwindles. If a galaxy is left alone – isolated – then the gas accumulated in the disc will slowly drain, transforming into the stellar population in a fairly predictable way: the star-formation rate drops as the fuel is used up over timescales of billions of years. As we have seen, if a galaxy merger occurs – as in the Antennae Galaxies – then a starburst can be trigged that causes the galaxy star-formation rate to increase suddenly in an event that might last several hundreds of millions of years. During this time, the gas gets used up more quickly.

Competing with the consumption of gas into stars, which is driven by the insatiable gravity, we also have the effects of stellar and black hole feedback, which, remember, are mechanisms that regulate growth by preventing too much gas from forming stars too quickly. This regulation prevents the universe from being awash with large numbers of very massive galaxies, and has allowed the evolution in the star-formation rate to take the form of a gentle hill rather than a sheer cliff. Again, the nature of feedback and its efficacy vary from galaxy to galaxy in a way that can be linked to a galaxy's mass (it is far easier to remove gas from a low-mass galaxy like a dwarf than it is from a massive one, for example, because the gravitational grip holding the gas in the latter is far stronger).

Galaxies are not just limited to the gas that gets trapped within them at the time of their initial formation. New gas can be accreted over time, sucked in from intergalactic space. We say this gas is gravitationally 'cooling' within the dark-matter haloes because it loses gravitational potential energy as it goes from a dynamically 'hot' state to a dynamically 'cool' state under the grip of gravity. This top-up of star fuel means that the rate of decline in the gas reservoir is not as strong as it might otherwise have been if no new gas was accreted. However, the amount of gas that accumulates in this way also varies from galaxy to galaxy (again, the rate depending largely on the mass of the galaxy). So, different galaxies go through different star-formation histories. In the end, consumption of gas in star formation is the winning factor, since not only is the replenishment of the gas reservoirs from intergalactic space declining over time, but in some cases feedback and environmental factors (such as the assembly of galaxy clusters) also serve to prevent gas accretion and new star formation. Combined, the result is the observed slowdown of the overall rate of galaxy growth over time.

What I've presented is a very simplistic, broad-brush picture of galaxy evolution, and the details of the processes outlined above are still being worked out and studied. The main point is that galaxies don't all undergo the same evolution – their individual star-formation histories are determined by a combination of intrinsic parameters, such as their total mass, as well as their local environment. The local environment of a galaxy is a very important factor; it can influence the rate of galaxy–galaxy mergers and interactions, and also result in other external processes that can affect galactic properties. The most extreme environments, as we have seen, are clusters. Galaxies in clusters are subject to a wide range of astrophysical processes that don't occur else-where: ram-pressure stripping, gravitational tidal 'harassment', gas 'starvation' and 'strangulation'. This rather dubious nomenclature reflects the intensity of some of these effects in altering a galaxy. 'Harassment', for example, refers to the multiple high-speed passes between galaxies in crowded clusters, which happen at relative velocities too high for merging, but cause gravita-tional disturbances in the galaxies as they orbit within the cluster gravitational potential well. Over time, this can reconfigure the stellar distribution in the galaxies within clusters, morphologically transforming them.

At any given epoch, the rate of galaxy growth also varies as a function of local environment. Today clusters are the regions where almost no new stars are being formed; most of the activity is happening in environments such as the one the Milky Way resides in. The evolution of the global star-formation rate applies to all galaxies, however, so that, while a distant cluster seen 5 billion years ago might have a lower star-formation rate than the surround-ing field at the same epoch, the average rate of star formation in that cluster is higher than is observed in a cluster of comparable mass today. So, in the past, clusters, groups and single galaxies in the average-density field all had higher star-formation rates than those environments today. Understanding exactly how the evolution varies between environments is a key area of research; measuring how galaxies have evolved differently in different envi-ronments (for example, studying some of the cluster processes described above) allows us to link the physics of galaxy growth to growth of the large-scale distribution of dark matter in the universe.

The shape of the evolution of the global star-formation rate tells us that to study the main epoch of formation of the most massive galaxies *today*, we have to look to the most intensely star-forming galaxies in the distant (early) universe. This is what we're doing with SCUBA-2, since it can detect those youthful starbursts that are the progenitors of today's massive galaxies (like the ellipticals). I'm part of a survey called the JCMT Cosmology Legacy Survey, the biggest survey to be conducted with the SCUBA-2 camera. Simply put, the goal is to observe (relatively) big areas of the sky and identify large numbers of distant submillimetre-emitting galaxies, called SMGs. Often these SMGs are extremely faint or even undetected at other wavelengths, particularly the optical bands, but they shine through in the submillimetre bands because of their vigorous but dust-enshrouded activity. Since the resolution of SCUBA-2 is far poorer than can be achieved with optical light, the images produced are not as 'pretty' as, say, the HST, or even a small optical telescope that you could buy yourself.

The reason for the low angular resolution comes from the fact that we're using a much longer wavelength than optical light. The angular resolution of a telescope (how sharply we can make images) is given, roughly, by the wavelength of light you're using, divided by the size of the dish or mirror you're using to collect it. For the JCMT, which is a 15-metre-wide dish, this is about 8 arcseconds at 450 microns and twice that at 850 microns – the two wavelengths that SCUBA-2 sees. For comparison, the HST has a resolution that is of the order of a tenth of 1 arcsecond, so it can see very fine detail (and we all know how spectacular the images produced by HST are), but only in the visible and near-infrared bands. So what we end up with from SCUBA-2 is an image that, where there is a bright SMG, contains just a bright blob of pixels. We cannot resolve spatial details in the galaxy. On the face if it, this is not very exciting to the general public, until you come to the realization that these blobs represent some of the most extreme powerhouses of the universe, churning out the equivalent of up to 1,000 Suns in new stars every year. And the great thing about the submillimetre bands is that we can exploit a rather useful quirk of nature to easily detect active galaxies in the *very* distant universe.

We mentioned earlier that the spectrum of the dust emission from galaxies is shaped like a black body, with a peak in emission at a wavelength of around 100–200 microns. At wavelengths beyond that peak, out to around 1 millimetre or so, the amount of energy emitted by the galaxy declines smoothly. The submillimetre bands we are using are right in the middle of this slope. As we look towards distant galaxies, however, the observed spectrum gets shifted to longer wavelengths because of the redshift, meaning that fixed SCUBA-2 bands see the dust emission closer to the intrinsic peak in the spectrum – the peak of the thermal dust emission has been redshifted closer to the observing bands. Of course, as we move a galaxy to further distances, its flux gets fainter at all wavelengths of light. However, because the dust spectrum beyond the peak is declining with increasing wavelength, the dimming effect that goes with increasing redshift is compensated for by the fact that SCUBA-2 is probing a part of the spectrum where the *rest-frame* emission is brighter.

This means that at a fixed luminosity, a galaxy like a ULIRG has roughly the same observed brightness across a vast swathe of cosmic history. This is like someone holding a candle in front of you and then walking away into the distance, but the candle not getting any dimmer. The practical upshot is that we can chase these galaxies out to far higher redshifts than we could do in, say, the optical or radio bands, where the shape of the galaxy spectrum does not allow for this trick. With the SCUBA-2 Cosmology Legacy Survey, we can measure star formation in galaxies potentially right back to a time when the universe was just 500,000 years old, close to the time when the very first galaxies formed.

The star-formation rate is one of the main observables we can track in bulk over cosmic time. The other main important observable is the stellar mass of galaxies. Armed with a redshift (or a good guess at it), so that we can convert the observed flux to total luminosity, the stellar mass of a galaxy can be estimated by measuring the total amount of optical and near-infrared light from a galaxy, since this emission comes mainly from the stars, with the total number of photons emitted proportional to the number of stars. Actually it's slightly more complicated than this, since, when viewed as a whole, galaxies contain a range of stellar types of different ages, releasing different numbers of photons at each wavelength (young, massive stars dominating the blue light and old, lower-mass stars dominating the red). Provided we have some idea of the stellar initial mass function, describing the distribution in mass of a given stellar population, and the average age of the stellar population (which tells us what the current stellar distribution should be, given the initial mass function), and of course an estimate of how much light we're missing because of dust extinction, we can estimate the total mass in stars in any galaxy.

When combined, the stellar mass and star-formation rates of galaxies reveal another clue to galaxy evolution. If we break up the galaxies into different divisions of stellar mass, then look at how the average star-formation rates have changed over time in each of those groups, we see that the peak of activity depends on the galaxy mass. Although the peak of activity for galaxies *on average* was 8 to 10 billion years ago, the peak epoch of star formation for the most massive galaxies occurred earlier in cosmic history than for the less massive galaxies. The name given to this observation is 'downsizing', which describes the idea that the bulk of stellar mass growth in the universe occurs in less massive systems over time.

This chimes in nicely with one of the observations we have discussed before: that the most massive galaxies in the universe today – elliptical galaxies in the centres of clusters – are also among the oldest galaxies in the universe, and that the peak of their growth was early. It's clear that there is a strong link between galaxy evolution and mass, but there is a delicate subtlety here: clearly there is a link between star-formation history and galaxy mass, but there is also a strong link between galaxy mass and environment: the most massive galaxies live in the densest nodes of the cosmic web. One wonders how much the link between the history of galaxy growth and galaxy mass is to do with 'local' physics relating to the galaxy itself, or physics related to the growth of the structures it lives in (environmental conditioning, if you like). More succinctly, if we were to track the evolution of two galaxies of the same mass, but in different environments, would we see the same thing? This 'nature versus nurture' question of galaxy evolution dogs astronomers. It's very hard to answer, and in fact, at a deep level, these are two sides of the same coin.

The role of galactic habitat

All structure in the universe has grown over time from what were originally tiny – quantum – perturbations in the distribution of matter (both dark matter and normal matter). Dense environments today, such as clusters, have *always* been dense environments relative to the average matter distribution. If we were to visit the site of, say, the Coma cluster, the most massive local cluster, just shortly after the Big Bang, we would find no stars or galaxies yet, but the density of matter here would be slightly higher than everywhere else nearby. As gravity is solely an attractive force and depends simply on the abundance of mass, dense regions – like our natal cluster – collapse (that is, become denser, accrete more matter and thus get more massive) sooner than other regions. The gas within these regions is able to collect and condense down into what will become proto-galaxies (not galaxies as we would recognize them yet) a bit sooner than everywhere else. Thus galaxies born in dense environments might have a kind of head start compared with other galaxies.

Just as a 'proto-galaxy' does not resemble a local galaxy, so a 'proto-cluster' – the dense region that will eventually become a rich cluster – does not resemble something like Coma or Virgo. Proto-clusters are more like a loosely bound collection of young galaxies and gas, gradually collapsing into a single, gravitationally bound structure. Importantly, the environmental conditions in this young cluster do not yet affect the evolution of galaxies in the wide variety of ways we see in dense, massive clusters today, which are the descendant environments of proto-clusters seen earlier in the history of the universe. For example, ram-pressure stripping can only occur if the cluster is filled with a hot, dense plasma, and this takes some time to form, as the cluster grows to form a deep gravitational potential. However, any proto-galaxies that form close to the peak in the matter density (the deepest part of the potential well), will remain there indefinitely, so the destiny of the galaxies that will become massive ellipticals in the centres of environments like the Coma cluster was in some sense imprinted on them by the environmental conditions of their birth, which in turn was randomly determined by those quantum fluctuations in the distribution of matter shortly after the Big Bang.

Clearly, for some galaxies, conditioning within their environment has had a profound impact on their development, even at late evolutionary times. Again, we look to clusters as an example. Clusters of galaxies continuously grow over time, accreting more and more matter. Part of this accretion is in the form of individual galaxies and groups of galaxies in the vicinity, which are pulled in through gravity. Once a massive cluster is properly established, then environmental effects play a major role in these 'infall' galaxies' development. The most important of these is to shut down star formation and cause morphological transformation as the galaxies traverse the cluster environment. My own research has touched upon this topic.

If we look at clusters today, we see that their cores contain a population of mainly passive elliptical and lenticular (S0) galaxies. If we observe clusters of galaxies at higher redshift, so that they are seen about 5 billion years ago (when the solar system was forming, incidentally), we still see the population of ellipticals, but the S0s are absent, or at least present in far fewer numbers. Where are they? The answer is that the S0s represent a population of galaxies that has accumulated in the cores of rich clusters over the past 5 billion years of cosmic history. One theory is that the S0s are the descendants of what were once large spiral galaxies that fell into the cluster and subsequently had their star formation stopped, either through ram-pressure stripping of gas, or through starvation of gas from being immersed in the harsh, hot cluster atmosphere. Over time these galaxies congregated in the core of the cluster – the bottom of the gravitational well – where they 'passively' evolved.

The mystery was that optical studies had not found a population of large spiral galaxies with star-formation rates high enough to turn into an S0. You see, S0s are really rather massive galaxies with large stellar bulges. To turn a typical spiral galaxy into an So requires additional stellar mass growth, especially in the bulge region. Although the abundance of star-forming spiral galaxies in clusters does increase as we look back in time, compensating for the absence of the Sos, it appeared from optical light studies that these galaxies lacked the star-forming punch to evolve into the Sos. Some research I was involved in aimed to address this problem: we undertook a search for star-forming galaxies in distant clusters that might be obscured by dust, so their star-formation rates had previously been underestimated. Our experiment made use of the mid-infrared imaging capabilities of Spitzer to make maps of several big clusters at a redshift of about 0.5 (seen around 5 billion years in the past). The goal was simple: search for galaxies with bright infrared emission and therefore high star-formation rates, potentially revealing a population of galaxies that are making the transition from spiral to lenticular.

Our experiment was a success. We discovered a population of luminous infrared galaxies (LIRGs, about a factor of ten less luminous than the ULIRGs we have met) forming stars at rates that were previously drastically underestimated. These galaxies were present in sufficient numbers and forming enough stars to suggest that they could easily build up the extra stellar mass required to form an So. The galaxies were mainly on the outskirts of the cluster, quite far away from the most serious ram pressure effects, so a lot of stellar mass could be assembled without environmental hindrance. Once deeper inside the cluster – we argued – the harsh cluster environment would prevent further star formation (a kind of cosmic preserving bottle); the disc would fade and the spiral arms would disappear, and the bulge would be bigger. Everything you need to turn a spiral into an So.

A couple of years later, we returned to the sample. We wanted to ask how much gas the Spitzer-identified galaxies, the putative progenitors of lenticulars, have. Measuring the star-formation rate is all well and good, but it's important to also have a handle on the amount of gas the galaxies have left for further star formation; the star-formation rate is an instantaneous measurement of what's happening to the galaxy right now. Was there really *enough* gas in the galactic tank to form the stars in an So? We managed to detect carbon monoxide in a sample of five of the galaxies in one of the clusters in our study for which we had measured the infrared properties very well. The total mass of gas in the galaxies, derived from the carbon monoxide luminosities, is some 10 billion times the mass of the Sun. This is unburned fuel: we knew that the galaxies *already* had rather large stellar masses (the optical and near-infrared data suggested masses of tens of billions of solar masses), but the gas observations confirmed that there was plenty of raw material to build sufficient additional stellar mass to match that of a typical So.

This subject is still part of my research. I'm currently trying to study these galaxies in more detail, to learn more about their physics. One of the major efforts is to try to obtain star-formation and gas observations at higher spatial resolution in order to examine *where* in the galaxies the stellar mass is being built. Is it all occurring in the bulge region, as we expect, or is it spread throughout the disc? It takes time to accumulate and analyse the data to answer questions like these, but this is one of the most exciting things about being a scientist – the feeling that you're on a path of discovery, each step finding out something new about nature that no one has ever known.

We've talked about the evolution of galaxies, and specifically the importance of infrared and submillimetre observations (the perfect example of why it's vital to have a multi-wavelength view of the universe). Now let's examine some of the techniques we can use to better explore galaxies in the distant universe.

Gravitational windows on the past

As I hope I have made clear throughout, astronomers are always battling with the ratio of signal to noise. The light – the flux – that falls on Earth from the most distant galaxies is tiny, so our view of the distant universe becomes increasingly uncertain as that signal becomes similar in size to the noise in our measurements. The brightest systems, such as quasars and intensely star-forming galaxies, are the easiest to spot and to measure, so the record-breaking galaxies (for distance) will always tend to be these extreme systems. This is what we call a selection effect; it's not that these extreme galaxies are the only galaxies out there, it's just that they are the easiest to detect. Normal galaxies, those like our Milky Way, are hard to spot in the distant universe. Luckily, nature has handed us a few tricks that allow us to see further. One of the most effective, and most remarkable, techniques in modern extragalactic astronomy takes advantage of a natural effect called gravitational lensing.

One of the predictions of Einstein's general theory of relativity, which describes the force of gravity through the contortion of space-time, is that a photon passing near to a large mass will be deflected due to the distortion of the space-time in its vicinity. The classic two-dimensional illustration of the effect is a bowling ball placed on a rubber sheet, which creates a deep dimple. If you roll a ball along the surface of this sheet and view its motion from above, you will see that the ball's path will stray from a straight line as it tracks around the depression in the sheet caused by the bowling ball. The same thing happens with light that passes by large masses like galaxies and clusters. The effect is called gravitational lensing, and, like a magnifying glass, we can use it to amplify the flux from distant galaxies.

Experimentally, the phenomenon of gravitational lensing was first demonstrated in 1919 by physics hero Sir Arthur Eddington, shortly after Einstein had published the theory itself. During a total solar eclipse, Eddington measured the position on the sky of a certain bright star close (in angular separation, not physical distance) to the Sun. The position of the same star, when observed at another point in the year when it was far from the Sun, revealed a change in position exactly in accordance with what would be expected if the mass of the Sun was deflecting the path of light rays passing close to it, as predicted by general relativity. This is one of the most exquisite astronomical observations of all time.

The mass of the Sun is large to us, but it is miniscule in astronomical terms (after all, a solar mass is our basic unit of describing mass in galaxies, like grams in a bag of flour). What about the gravitational lensing caused by the most massive systems in the universe: galaxy clusters? When very deep, long-exposure images of galaxy clusters were obtained it became clear that some of them appeared to contain blue, arc-like features surrounding the dense core of red elliptical galaxies. These blue arcs are not physically part of the cluster, but very distant galaxies that happen to lie along the same line of sight to the cluster. The light emitted by those distant galaxies has passed through the massive cluster on its way to our telescopes, and been deflected in an analogous way to light passing through a glass lens.

Galaxies are not single points of light, so what is intrinsically a disc shape, say, gets stretched out – different parts of the galaxy are bent and deflected by slightly different amounts depending on the distribution of mass in the

A dense cluster of galaxies, Abell 1689 (from George O. Abell's catalogue of rich clusters of galaxies). Clusters are among the most massive gravitationally bound structures in the universe, and the galaxies in the cores of clusters like this are dominated by the elliptical type: massive, old galaxies with little ongoing star formation. The galaxies formed early on in the history of the universe, birthed when the largest density fluctuations in the initial matter field grew into massive structures like this. Abell 1689 beautifully reveals the signature of the gravitational warping of space-time predicted by Einstein's general theory of relativity: around the core of the cluster can be seen 'gravitational arcs' – images of more distant galaxies along the line of sight, the light from which has been bent and distorted by the intervening matter. Not only can gravitational lensing be used to study very distant galaxies in great detail (since they are magnified), but the pattern of distortions can be used to reconstruct the total mass of the cluster, providing evidence for the existence of a dark-matter component.

lens (the cluster, in our example). Like a magnifying glass, gravitational lensing also *amplifies* the light of distant galaxies, making them appear brighter than the galaxy would be if the foreground lens were not in the way. Lensing is a purely gravitational effect and depends on the total mass present in the system, light and dark. Therefore gravitational lensing provides another method from which to infer the presence, and indeed examine the distribution, of dark matter in clusters and galaxies. If we subtract the mass that is in the form of visible stars in galaxies, and the gas between the galaxies (which is a not insignificant part of the mass budget of a cluster), and compare that to the 'lensing mass', we find the excess mass that is the signature of the presence of dark matter.

The long, orange-blue streak in this view of the cluster of galaxies Abell 370 is the distorted image of a distant galaxy that has been gravitationally lensed by the massive foreground cluster (other lensed features are also seen as linear features around the bright elliptical galaxies). The shape of the lensed image is determined by the distribution of the matter in the cluster (both normal matter and dark matter). The magnification of brightness and stretching of the image of the distant galaxy by the foreground cluster allows us to examine the properties of this galaxy in far greater detail than would otherwise be possible.

The bright arcs seen around the cores of galaxy clusters are examples of what is called 'strong' lensing, because they represent light rays that are significantly distorted by the cluster mass. Although we don't see these bright, strongly lensed arcs at large angular separations from cluster cores, the background galaxies at large radii from cluster centres are still affected by the foreground mass, but to a lesser extent. These small distortions are almost imperceptible: the shapes of the galaxies are only very slightly distorted and the observed fluxes only slightly magnified. The effect is so subtle that it can only be seen in a statistical analysis of the shapes and fluxes of many galaxies. Luckily there are so many galaxies that this *can* be done in deep-imaging surveys. This is called 'weak' lensing.

We can use strong lensing to study distant galaxies in more detail than is otherwise possible, exploiting the fact that their observed flux is boosted, and so easier for us to detect, increasing the signal-to-noise in observations like spectroscopy. The distorting effect of the lens also stretches the apparent size of the distant galaxies, and this allows us to see things on a smaller physical scale compared to their unlensed counterparts. Gravitational lensing therefore offers valuable opportunities to study very distant galaxies in great detail, working best when the distance to the galaxy being lensed is about twice the distance from us to the lens itself.

There is a catch in using lensing to study distant galaxies. Actually there are several catches. First, we can't move galaxy clusters around, so we're limited to observations of the relatively small number of distant galaxies that fortunately are aligned along the line of sight. Clusters are pretty rare in the first place, and not all of them appear to act as strongly lensing systems. So our sample size of lensed galaxies is small compared with the myriad galaxies in the unlensed 'field'. Second, although the magnification and distortion of the images of those galaxies helps us, it also makes analysis slightly tricky because we have to reconstruct what those galaxies really look like in the 'source plane' – that is, if the cluster wasn't there, what would they look like? We can do this by constructing 'lens models' that attempt to model the distribution of mass in the lens, using the shapes and orientations of all of the distorted images of different galaxies around a given strong lensing cluster (a single galaxy can be lensed into multiple images, and there may be several independent galaxies lensed by the cluster). This requires very high-resolution imaging, and the Hubble Space Telescope has been the key instrument making this analysis possible, delivering the sharp imaging required to detect the lensed galaxies in the first place and also to perform the lens modelling.

Basically, clusters of galaxies can be treated as giant telescopes with 'mirrors' hundreds of thousands of parsecs in extent. Fascinatingly, the very structure of the universe gives us a leg-up when it comes to studying the galaxies within it. But lensing can only take us so far. The Holy Grail – or, perhaps better, the final frontier – of galaxy evolution studies is to peer right back into the universe's past to the time when galaxies first formed. This era is called the Epoch of Reionization.

The genesis of galaxies

Imagine spilling coffee on a cobbled floor: the liquid collects in the nooks and pits between the rounded, raised stones. Liquid that lands on the top of a cobble quickly drains off into the gaps between the stones. When on top of the cobble, a drop of coffee has some gravitational potential energy, but it can lose this by draining into the valleys. The gravitational potential energy turns into kinetic energy as the drop drains down the side of the cobble, and this kinetic energy dissipates once the drop settles to the lowest points, sloshing among the stones. The starting point for galaxy formation can be thought of as primordial gas settling into the 'cobbles' of the early universe.

After the formation of the universe in the Big Bang, there was no large-scale structure as we see today, and the (then small) cosmic volume was filled with a hot plasma of normal matter mixed in with a sea of dark matter. This normal matter represents the base material required to form everything we see around us today, comprising protons, neutrons and electrons. The dark matter served as a kind of 'skeleton' to which the normal matter could stick as gravity amplified small imperfections in the density distribution. After enough time had passed, and the universe had cooled sufficiently during its expansion, electrons could combine with the nuclei of the simple elements – mainly hydrogen and helium (and some deuterium and lithium). This is called the Epoch of Recombination, and represents a time when the matter in the universe went from being ionized (free electrons) to neutral (electrons bound to atoms via the electromagnetic force). The scene was set for the entire future of galaxy formation.

While this was happening, the small density fluctuations in the matter distribution started to attract more material: both dark matter and normal baryonic matter. This was the moment at which galaxies started to form: neutral gas pervading space started streaming into the over-densities growing in the matter field. Where the gas pooled into the first dark-matter haloes, proto-galaxies formed. After a critical point, where the gas density grew large enough to ignite nuclear fusion in the pristine (metal-free) gas, the proto-galaxies began to form stars. As soon as this happened, those first stars bathed the surrounding space with photons, illuminating the surrounding neutral gas. Some of these photons (the ultraviolet ones) are energetic enough to remove electrons from atoms of neutral hydrogen, reionizing them. It is thought that the growth of supermassive black holes, which started to form soon after the first stellar generation, also contributes to the progression of reionization, since they radiate energy while accreting matter. You can think of this process like the spread of a plague: bubbles of ionized gas inflating around the bright young galaxies, pervading nearly all space. This is why the era is called the Epoch of Reionization: the universe went from an initial state of complete ionization, through a neutral phase, then was ionized again as the first galaxies switched on.

Currently, the Epoch of Reionization is just beyond our observational grasp, but only by a little bit – we are very close to observing it properly, but are not quite there yet. In a few years, new radio surveys should deliver the data that will allow us to pinpoint this change in the universe (for we don't know exactly when reionization happened, or how long it took), because it will be possible to detect the signature of the emission of neutral hydrogen (the 21 cm line we have talked about) at very high redshift, where reionization is expected to happen (around a redshift of 10, or a few hundred thousand years after the Big Bang) as the line is redshifted into the low-frequency part of the radio spectrum. These observations are hard because this signal is weak. In the meantime, most of the research in this area is on the theoretical front: what do we *expect* to see, given current models? One of the major ingredients of modern astronomy is the use of powerful computer models and simulations to explore our understanding and develop hypotheses of how the universe works.

This is an image from the Millennium Simulation, a large computer simulation of the behaviour of Cold Dark Matter, represented by a very large number of particles. Given a cosmological model, the simulation tracks the evolution of dark matter from initial conditions. The present-day distribution of dark matter around a massive 'halo' (a cluster of galaxies) is shown here (the view spans nearly 100 million parsecs – a representative chunk of the universe), where brighter colours correspond to a higher dark matter density. Galaxies form within these haloes, with the central massive structure perhaps containing many individual galaxies (Coma would be an example of this). The distribution of matter in the cosmic web can clearly be seen, as can the 'hierarchy' of structure. Our current model of galaxy formation posits that galaxies as we know them form as primordial gas 'cools' within dark-matter haloes, forming stars. The exact details of galaxy formation are highly complex, involving a wide range of physics, and as yet, we do not understand the true nature of the dark matter itself.

FIVE

Models of the World

To make a dangerous generalization, there are two factions of astronomers: observers and theorists. Simply put, theorists spend most of their time constructing and testing models of the universe, or bits of it (like galaxies), and speculating how a particular astrophysical process works from first principles. Where we lack complete knowledge of how nature works (such as how a galaxy forms), a theorist's job is to come up with a plausible model to explain it. These models are compared with data to see if they work. That grist comes from observers.

Observers are empiricists, and I class myself in this bloc. Using measuring devices (in our case telescopes and all their accoutrements, such as cameras and spectrographs) we obtain data to observe phenomena, then interpret that data in the framework of our current 'World Model', the cosmological paradigm that describes the universe as a whole. Of course, there is overlap between the two clans (there has to be, in order for progression to be made via the scientific process of testing models against evidence), and there are those that can bridge the two camps, exploiting the advantages of both approaches. But there has traditionally been rivalry, verging on class warfare, between the observers and theorists that dates back to the time of Newton and Flamsteed in the seventeenth century. Back then, our World Model of the universe was focused on the celestial mechanics of the solar system, with the motions of the planets and comets observed by people like Flamsteed, who provided the data with which Newton could produce his masterpiece work on gravity. Without the observations, this great theoretical leap in our understanding of gravity could not have occurred. The same rule applies today.

Three hundred and fifty years later, our World Model is called Lambda-CDM (also referred to as the concordance cosmology), and is incomplete, in that we don't fully understand its main components. Lambda is the symbol for dark energy (the origin of the symbol is from Einstein's equations, representing a term called the 'cosmological constant'; he thought it might be a mathematical quirk, and at the time referred to this inconvenient term as his 'greatest blunder'). Dark energy is the name given to the mechanism responsible for the observed acceleration of the expansion of the universe, as evidenced by the brightness of distant supernovae. Like dark matter, it's 'dark' because we don't know exactly what it is (although we have ideas), and in terms of the total energy density budget of the universe, it appears to dominate over dark and normal matter combined.

We're not going to talk about dark energy here, because as far as individual galaxies go, it's not relevant to our discussion, and in terms of the history of the universe, dark energy has only really kicked in as a significant influence fairly recently. Assuming the acceleration continues, this will be more important in the future evolution of the large-scale structure of the universe. CDM stands for Cold Dark Matter, the other main mass component in the universe and, as we have seen, an important ingredient in the formation, structure and distribution of galaxies. The 'cold' in CDM refers to the idea that the particles that make up the dark matter move slowly compared with the speed of light. There are other models in which the dark matter is 'warm', and these models produce different predictions for the evolution of structure in the universe. Again, this is something we can test against observation. Currently CDM is favoured by most, but WDM also has its proponents.

We haven't yet directly detected dark matter in experiments, despite the fact that, by mass, it overwhelms normal baryonic matter by about five to one. The problem is simple: dark matter doesn't appear to interact strongly with other matter in any way apart from gravity, and we have to look on astronomical scales to see this (the rotation curves of galaxies, or gravitational lensing, for example). If dark matter is made from particles called WIMPs (Weakly Interacting Massive Particles), then we might expect to see – very occasionally – the recoil of a particle of normal matter when two of these particles collide (the 'Weakly' in WIMP implies that an interaction between normal and dark matter *can* occur, but those events are rare). There are experiments ongoing that aim to search for this very effect.

An example of a dark-matter detection experiment was the ZEPLIN-III detector, which consisted of about 12 kilograms of liquid xenon with a bit of gaseous xenon on top (xenon is one of the noble, inert gases). In this liquid photomultiplier tubes were immersed, with the idea being that the photomultipliers would detect and amplify any signature of the recoils of xenon atoms when a WIMP collides with them, which produces a brief, tiny burst of light. To reduce the contamination of the signal from other particles passing through the xenon whose passage might trigger the detectors, the detector was placed over 1 kilometre underground in the Boulby potash mine (located in Britain's North York Moors), with the thick layer of rock shielding the detector from any contaminating signal. One example of contamination is cosmic rays – high-energy radiation that is constantly raining down on our heads from various energetic astrophysical processes, like supernovae. The flux of cosmic rays is significantly damped by a few hundred metres of rock, making a mine the perfect location for experiments like ZEPLIN-III. There has yet to be really convincing observational evidence for direct dark matter detection, and dark matter might not be made of WIMPs, but the search continues.

Despite its uncertainties, Lambda-CDM represents our best model of the universe. I can understand why the public might be sceptical about the actual existence of something like dark matter – it is thought to be everywhere (although the density changes from place to place) but has no obvious impact on our everyday lives. But when you look on cosmic scales, there is indirect evidence for a dark-matter component everywhere: in the rotation curves of galaxies and the motions of stars, and in the gravitational lensing of light. Regardless of precisely what dark matter is, and the exact nature of dark energy, our current model that describes them seems to do a good job of predicting many things about the universe that agree well with observation. There *are* problems with the model, but that's not entirely surprising: the whole purpose of our research is to build a model that can be refined over time with the accumulation of new knowledge. If we found strong observational evidence that could not be explained by the model, or showed that the model was wrong, then it would be thrown out and we would start again. An example of this is the paradigm shift that occurred when the model of the universe starting with a hot Big Bang took over from the 'Steady State' theory.

Origins in a maelstrom

The Steady State theory, championed by the likes of Fred Hoyle, Hermann Bondi and Thomas Gold in the mid-twentieth century (all were, and remain, highly respected astronomers and cosmologists), described a universe that had no beginning but had always existed. Galaxies are still in motion in the Steady State model (as evidenced by the recession of galaxies relative to the Earth, discovered earlier in the century), but the universe keeps a constant density (when averaged over large volumes), by allowing new galaxies to come into existence as the cosmos expands. The two main pieces of evidence that discount the Steady State theory are the observation of the Cosmic Microwave Background, a pervasive radiation field that implies a hot origin to the universe, and – pertinent to our story – the fact that galaxies at high redshifts start to exhibit different properties from those locally. For example, there appear to be many more quasars in the distant, early universe than exist today. This implies that the galaxy population is changing over time. As it happens, the higher abundance of quasars at earlier times in the history of the universe is linked to the fact that the rate of activity – both star forma-tion and black-hole growth – was higher in the past than it is today, indicating a progressive change in the galaxy population.

With mounting evidence, the Steady State theory fell by the wayside, and with its demise we could cross off one of the theories of how the universe works. This was not a wasted effort, by any means; science is driven by the empirical testing of hypotheses, and the Steady State simply didn't stack up against observations. The Hot Big Bang theory took over, describing the instantaneous formation of space and time from a single point at a finite moment in the past. The Lambda-CDM model describes the contents of this universe and its geometry and evolution. It's not perfect, and astronomers do recognize this (well, some more than others – it's easy to become dogmatic in this business). For example, there are problems with how the theory works at very, very early times – right after the Big Bang – in describing the mechanics of how the universe expanded so rapidly. Similarly we have no testable theory of how the Big Bang happened in the first place, or what came before, or whether there are *other* universes. That's another story. For now, astronomers like me use Lambda-CDM as a context or a framework within which to help interpret observations, and – more importantly – as a model that can be held up to scrutiny. As I have mentioned before, our current World Model is remarkably successful at explaining a wide range of phenomena, so we're probably on the right track. On the other hand, on the scale of individual galaxies and their internal workings, good old-fashioned physics is involved. On their own, many of the principles of the physical processes operating in galaxies are fairly well understood. The problem comes when we try to understand how all the many different physical processes operate together within galaxies. That's when things get tricky: we have to work piece by piece, seeing how it all fits.

The thermodynamic properties of the early universe are such that there is a fundamental horizon beyond which, as observers, we cannot see. As we have seen, the fates of dark and normal baryonic matter have been entwined from the start, and shortly after the formation of the universe, both dark matter (whatever it is) and baryonic matter were distributed smoothly in a hot mixture. We cannot directly observe this epoch because photons moving within the hot plasma were effectively trapped by being constantly scattered off the charged baryonic particles. This constant scattering meant that they didn't get the chance to free-stream across the universe in the same way that light from distant galaxies does, relatively unimpeded by intervening matter. But once the universe expanded and cooled enough for electrons to combine with protons to form the first atoms, neutralizing the universe, this scattering all but stopped, and the photons – radiation from the Big Bang itself – were released like horses at the start of the Grand National, streaming virtually unmolested to us across an ever-expanding universe on a journey of nearly 14 billion years. The point at which these photons were released is called the Epoch of Recombination, allowing photons to be released from the 'Surface of Last Scattering'. This is the most distant (or if you wish, earliest) thing we can see. This surface, or rather the emission from it, that permeates the universe, is called the Cosmic Microwave Background (CMB).

The CMB is a nearly uniform rain of light that, with the expansion of the universe that has occurred throughout its journey, has been redshifted to microwave wavelengths, and appears to emanate from every direction in the sky (although the signal from the CMB is dwarfed by the microwave emission from the Milky Way itself). The spectrum of the CMB is a near-perfect black body, representing thermal emission of radiation with a characteristic spectral distribution similar to the shape of the infrared dust emission of galaxies that we have encountered before. The peak of the spectrum corresponds with an average temperature of just 2.73 degrees above absolute zero, and this represents the ambient temperature of space – the residual heat of the Big Bang.

The CMB, when mapped over the entire sky by satellites like the Cosmic Microwave Background Explorer (COBE), the Wilkinson Microwave Anisotropy Probe (WMAP) and, most recently, Planck, is not smooth. There are ripples in the temperature which, although small (the variations are of the order one part in 100,000), are of monumental significance to the story of galaxy evolution. These ripples in temperature represent the density fluctuations that were present in the hot soup of particles just a few hundred thousand years after the Big Bang. These are the cobble stones. The fluctuations in the CMB are the sign that baryons were beginning to settle within the high-density regions that were growing from even earlier quantum perturbations in the density of matter as the universe rapidly expanded from a single point. The details of the distribution of galaxies we see around us today were encoded at this time, as baryons flowed into, and then themselves helped amplify, these gravitational furrows. The CMB is a photograph of this time: a snapshot of the universe when galaxies were *just* starting to form. Our ability to map the CMB in the detail we can is one of the crowning achievements of observational cosmology.

Missing baryons in the skeleton of the universe

I think of dark matter as the hidden skeleton that is laced with the visible matter – gas and galaxies – that we can see. The largest surveys reveal that galaxies are distributed in filaments, clusters and groups linked together in a web-like network of large-scale structures, like the pattern left on the side of a glass after a frothy pint. In our current model, galaxies are tracing out this unseen network of dark matter, just as the luminous glow of street and house lights betrays the locations of roads, towns and cities on Earth when viewed at night from space.

The gravitational influence of dark matter has helped shaped galaxies, corralling the baryons into ordered structures like the Milky Way. But there is another interesting side to the evolution of baryons. We have talked about it throughout this book, and you are probably already aware that normal matter makes up only a fraction of the total mass of the 'material' universe, and that the rest of the mass is dark. A lesser-known issue is that only a small fraction of those baryons, which represent such a trifling fraction of the total mass, is actually in galaxies at all. We know how many baryons there should be in the universe in total from studies of the CMB (the statistical distribution of fluctuations in the temperature of the CMB background is encoded with lots of information about the conditions of the universe just before galaxies properly formed, including the 'baryon fraction'), and also in measurements of the abundance of primordial elements like helium, deuterium and lithium. These, the lightest elements, were able to form shortly after the Big Bang in a process called nucleosynthesis and their relative abundance is controlled by the total baryon density relative to all matter.

We can add up all of the mass in galaxies from the starlight (visible and near-infrared light), the gas (radio and millimetre wave) and the dust (infrared). We can even add up the mass in baryons outside of galaxies, through the x-ray glow of the hot atmosphere in clusters of galaxies, and in the absorption lines of elements in extragalactic space that happen to be back-lit by bright quasars. But when we add all this up, we find we can account for fewer baryons than the total expected. The rest are missing, and this mystery has become known as the 'missing baryons problem'. It implies that we don't fully understand galaxy formation. Which of course, we don't, otherwise I would be out of a job.

As we look into the distant universe, we *can* find evidence of some of the baryons that are now missing. Again, using quasar spectra, we can search for the imprints made from clumps of neutral hydrogen gas in and around galaxies, as well as floating around intergalactic space. A single cloud of neutral gas can absorb some of the quasar light, leaving an absorption line at a specific wavelength corresponding to the redshift of that cloud. Light from quasars that are more distant has to pass through more of intergalactic space and this can intercept many clouds along the way, introducing many absorption lines in the quasar spectrum. So many absorption lines, all at different wavelengths, can build up in the quasar spectra that the network of gas clouds is called the 'Lyman-alpha forest'. Lyman-alpha is the name of the absorption line of hydrogen in question; it is the principle transition in the 'Lyman' series of the hydrogen atom, relating to electrons in the lowest energy level of the system. By measuring the abundance of these clouds of neutral hydrogen, and their masses (determined by the strength of the absorption), along with the mass in stars formed in galaxies at the same time, we can measure the total number of baryons earlier in the history of the universe.

As we look back to earlier times, we can account for more of the baryons that make up the theoretical total than we can today. Somewhere in between then and now, they have been lost; our best guess is that, over time, most of the baryons never made it into galaxies, or at least did not form cold gas or stars.

The missing baryons issue stems from a simple shortcoming: we still don't fully understand the baryon cycle, the flow of gas in and out of galaxies. We already know that there is a lot of gas in intergalactic space, the most obvious being that in clusters, where the intergalactic gas is hot enough to glow with x-rays. Clusters stick out like a sore thumb in x-ray images of the sky; but most individual galaxies are not strong emitters of x-ray radiation, and when they are, as in the case of quasars, for example, that emission is very compact. The sheer sizes of clusters give rise to extended x-ray emission, engulfing the comparatively minute galaxies immersed within this hot atmosphere. The total x-ray luminosity of the cluster can be converted to a total gas mass. But we can't detect baryons if they don't help us out in this way, by either emitting or obscuring radiation. One theory is that the missing baryons are simply in a state we have trouble detecting – a warmish gas that is hotter than the gas within galaxies, but tepid compared with the tens of millions of degrees intracluster medium. As such, this gas is not a prolific emitter of x-rays, or any other radiation that we can easily detect. That is our Achilles heel.

Where is this matter, if not in galaxies? Well, the answer comes from the dark-matter architecture, the large-scale structure that galaxies are embedded in. It is the filaments between clusters, containing many of the galaxies, that are thought to harbour a large fraction of the missing baryons. Just as clusters represent gigantic dark-matter haloes within which galaxies reside, the galaxies in filaments are embedded within haloes that mingle to form an interconnected mesh of dark mass, which has grown more massive over time.

This mass is sufficient to attract, accelerate and in the process heat intergalactic and primordial gas. The gas is heated to fairly high temperatures of a few hundred thousand degrees to a few million degrees, which is a phase too hot to collapse into galaxies and too cold to emit x-rays we can see. It is unused material: a vast reservoir from which new galaxies could form, or existing ones grow, but is held in limbo by gravity. This material is called the Warm-Hot Intergalactic Medium, or whim. Experiments to detect the whim rely on the same absorption-line technique we talked about before that allowed us to detect neutral gas in the spectra of distant quasars. The trick is to find a bright, distant quasar along the line of sight to a filamentary structure and obtain an *ultraviolet or x-ray* spectrum of it. If the light from the quasar passes through a dense patch of WHIM, then the light could get absorbed by strongly ionized elements present, such as oxygen. An oxygen atom that has had almost all of its electrons removed will absorb light at the high energies traced by ultraviolet and x-ray bands, and the detection of such a highly ionized element would signify the presence of a highly energized gaseous medium. X-ray absorption traces very hot gas, and ultraviolet absorption traces less hot but still warm gas. This has been seen in our local universe in environments like the 'Sculptor Wall'. Sculptor is a constellation, and in the direction of that constellation can be found a wall-like overdensity of galaxies that represents part of our local large-scale structure. X-ray spectra of a bright quasar behind the wall show a distinctive dip – an absorption line – at exactly the right wavelength expected for heavily ionized oxygen lingering between the galaxies in this dense structure.

Most of the WHIM is made from hydrogen; the oxygen and other heavy elements used in the absorption line work are simply trace contaminants. The only place where these heavy trace elements could have been created was inside stars inside galaxies. So, these 'pollutants' must somehow have escaped from the galaxies they were formed in, and now reside in this WHIM environment. This is further evidence of the cycling of gas in and out of galaxies: the baryon cycle. What about the bulk of the WHIM – how did it get there? This is best modelled in computer simulations, which are powerful tools in our understanding of the universe and its contents.

A toy universe

Astronomers nowadays use computers a lot. In my own research, most of my work involves the analysis of data from telescopes. From 'raw' images of the sky, what we aim to do is produce a calibrated, science-grade data product (say, a deep image or spectrum), then harvest this for interesting information. The processing required to do this relies on increasingly powerful hardware, since the sheer amount of data produced by telescopes is always growing, requiring larger digital storage facilities and more processing power and random access memory to deal with it. For theorists, on the other hand, the most important use of computers is to 'simulate' the universe, or at least a decent-sized chunk of it. Just like observers using telescopes, simulators are limited by hardware. There is always a want for faster machines, more memory and more processors, and to acquire these things as cheaply as possible. In the same way that observers want to see further and clearer, simulators always aim to perform bigger, better, higher-resolution simulations.

Arguably the most important types of simulation that we have of the large-scale properties of the universe are N-body simulations. The N in N-body stands for 'number of particles'. The simplest N-body simulation is called the two-body problem, and if you have a spare afternoon you can do this on a piece of paper. Draw a grid of lines on the paper – this is your model of the universe, restricted to two dimensions – and, like on a map, label the square cells you have defined by their positions on the grid. Now, pick two random cells and draw a dot in each. These are the two bodies of the two-body problem. We'll label them A and B. What we're going to do is model their evolution assuming that they are governed by the laws of physics. In this case, we'll just consider gravity.

We'll imagine the simplest case, where the particles A and B are initially at rest, and each has a unit mass. If we assume that gravity can just be described by Newton's description (the classical view), then each particle experiences a force, which is simply given by the product of its mass multiplied by a constant factor (called G, the universal gravitational constant – the exact value of it doesn't matter in this model), divided by the square of the distance between the particles. Each particle experiences this force, and is impelled to move, because the force causes an acceleration in the direction of the other particle; the size of the acceleration is equal to the strength of the force divided by the mass of the particle (one of Newton's laws of motion).

The next step is to get a new bit of paper, redraw the grid and then calculate the location of those particles assuming that some interval of time has passed. We can make that interval as long or short as we want, but the shorter it is, the more accurately we can track the positions of the particles. Then we repeat the process: calculate the force on each particle, the accelerations, add this to the current velocity of each particle, and so on. In this example the results are boring – the two particles are just mutually attracted and therefore accelerate towards each other, resulting in a steady state where the positions of the particles become locked together; our simulation doesn't know about any of the physics of the collisions between particles.

Things get more interesting when we add another particle, so it's now a 'three-body' problem. Since gravity operates between all objects with mass, we have to calculate the total force on each test particle given by the vector sum of the gravitational force between each pair, A – B, A – C, B – A, B – C, C – A and C – B. It's suddenly got a bit more complex in terms of the number of calculations we need to do to predict the evolution of the system. Now, instead of doing this in two dimensions, let's do it in three. And instead of three test particles, let's use millions. This is where supercomputers come in handy.

N-body cosmological simulations don't attempt to model the evolution of every particle in the universe. Far, far from it. Instead, a single particle might represent a rather large chunk of mass, but if your aim is to model the evolution of the large-scale structure of the universe, this coarse mass 'resolution' is fine, because it's acceptable to gloss over the fine structural details of, say, an individual galaxy, or a solar system. If you did want to simulate something like an individual galaxy in great detail, then with the equivalent number of particles you can, but at the cost of not simulating the rest of the universe, because you now have to represent a much smaller volume. The total number of particles that can be simulated depends on computer power, and clever algorithms, such as 'tree codes' and the 'particle mesh method,' have been developed to efficiently calculate the relevant forces acting on each particle without resorting to the 'brute force' method of simply looping over every one.

One of the most famous and successful N-body dark-matter simulations of recent years is called the Millennium Simulation, a project led by an international group of universities called the 'Virgo Consortium', headed by Durham University's Institute for Computational Cosmology in the United Kingdom, and the Max Planck Institute for Astrophysics in Garching, Germany. Many different research groups run their own simulations, but the Millennium Simulation is one of the most well known. The goal of the Millennium Simulation was to model the evolution of dark matter in a large chunk of a toy universe – a box of size 500 megaparsecs – using 10 billion particles to represent dark matter, with each particle having a mass of around 90 million times the mass of the Sun. So in this simulation, an individual galaxy might comprise a collection of 100 or more particles. A dark-matter halo, which might contain a galaxy, is defined as a clump of dark matter within which the density exceeds some threshold value, usually taken to be something like

200 times the average density of the universe. This division into haloes is a convenient way to describe structure in the universe, or at least the model universe.

At the time, the Millennium Simulation was the largest N-body simulation ever conducted. It ran for nearly a month of real time on 512 processor cores of an IBM supercomputer. This is equivalent to 350,000 hours, nearly four years, of CPU time. Millennium consumed 1,000 gigabytes of physical memory (typical desktop machines today have a few gigabytes of memory), performed nearly a billion billion floating point operations and churned out 20 terabytes of data. The whole point of Millennium was to see how a model universe (consisting of only dark matter) would evolve from initial conditions according to the input physics, which was our best guess given Lambda-CDM model parameters. How does structure form from the nearly

This image shows the same volume of the universe from the Millennium Simulation shown on page 224, but seen at an earlier epoch (that is, as it would appear at high redshift). Structures are in the process of collapse, and the central massive halo has not yet developed properly – it is currently a network of filaments and smaller haloes that are gradually taking shape. As extragalactic astronomers looking back in time, we aim to understand how the range of galaxies we actually see grew in relation to this underlying dark 'skeleton' of the universe.

smooth matter distribution at the start of the universe to the complex web at the present day? How do dark matter haloes grow? What is the distribution of the masses of dark-matter haloes and how does this evolve? The simulation gives us a way to visualize the theory and examine its predictions in a way not possible with pen and paper.

When we look at the evolution of dark matter in the Millennium Simulation, the development of a rich hierarchy of structure becomes clear: from a very smooth starting point, a complex environment – the cosmic web – forms. You can watch the collapse of matter at the sites that had slightly higher density initial conditions, growing over time through the accumulation of smaller clumps. In time, the largest haloes, representing the clusters, stand out as the densest nodes in a universe-pervading lattice of structure. Within and around the large haloes are smaller sub-haloes, forming a hierarchy that starts with dwarf satellites around bigger galaxies, which in turn cluster together into a larger and larger framework. The N-body simulations reveal the repeated merging of haloes: events that we see in the galaxy population around us as the cataclysmic collision of entire star systems, profoundly changing the history of those galaxies. These are commonplace, routine events in the model, just a natural part of the evolution of structure.

In the real universe, of course, we only see the baryons. We see the massive haloes of clusters filled with their hot gas, and inside them hundreds or thousands of luminous galaxies. N-body simulations can handle the dark matter that describes the framework fine, but what about those baryons we actually see as the luminous matter?

Baryons can also be simulated using particles, but this time, instead of just gravity acting on them, we have to include the extra physics: the particles have to be 'told' about the rules of thermodynamics, fluid mechanics and the transport of radiation, for instance. A technique for doing this is called 'smoothed-particle hydrodynamics' (SPH), which calculates the fluid properties at any point in a simulation grid from the contribution of many particles whose properties have been smoothed (averaged) out over their local volume. In cosmological 'hydro' simulations, one can track the evolution of a fluid – the primordial gas – and how it evolves alongside the dark matter. The baryon physics is very complex and therefore quite expensive to simulate in terms of processing power (often smaller volumes are tackled). Like all simulations, there is a resolution floor in SPH. When simulating a cosmological volume, there is not enough resolution to model the physics of, say, a star-forming cloud within an individual galaxy. It's possible to see how gas flowed into a clump of dark matter, reaching a high density, but after that we have to use a short cut to predict how many stars form, and at what rate. This is called sub-grid physics, because it requires assumptions about the evolution on a scale smaller than the simulation can 'see'.

When gas is added to these simulations, it is possible to track its co-evolution with the dark matter. As the dark-matter haloes grow from initial perturbations in the smooth-matter field, some of the gas flows into them, attracted by that same undiscerning force of gravity. We can watch the galaxies being born, analysing how the gas is being funnelled into the gravity well and how it is affected by processes such as star formation, supernovae and black hole growth. But the simulations show that a lot of the gas does *not* enter the haloes; it is attracted to and accelerates towards the large-scale filamentary structure that is also forming in the volume, which itself has significant attractive power.

Simulations suggest that, as this process happens, the gas is heated up. The amount of heating depends somewhat on the total gravitational energy of the system, which is why the gas that gets sucked into the dense clusters gets heated the most, up to x-ray-luminous temperatures. The gas flowing into the filaments only gets heated up to a few millions of degrees: the WHIM. That WHIM gas can only collapse into the galaxies forming within those filaments if it loses some of that energy, and this is what prevents a large fraction of the total mass of baryons in the universe from condensing into galaxies. Of course, there is a constant exchange – some gas *does* manage to cool into galaxies, providing a new supply of fuel for star formation. At the same time, however, gas is being ejected and energy is returned into the intergalactic medium from the galaxies themselves (both radiation from the stars and kinetic energy from outflows, as we saw in M82). So there is a continuous battle over these baryons, mediated by gravity and the competing forces of galactic feedback. X-ray and ultraviolet absorption line studies go some way to confirming the existence of an elusive baryonic component of the universe, but the observations are quite challenging, and there is only a limited number of elemental 'species' that can be used as probes, giving us a limited picture of the WHIM. Worse still, 'backlit' absorption line studies require something bright in the background against which we can see the contrast of the absorbing matter in the foreground. In most cases, these are usually distant, luminous quasars. The fortuitous alignments of distant quasars with dense parts of the

WHIM are rare, further limiting these studies to 'pencil beams' emanating from the Earth. This is an example of a model and simulation making a clear prediction about the evolution and distribution of gas in the universe that can be tested against observation. In the case of the detection of the WHIM, the observations are hard, requiring long exposures with space-based facilities, chiefly the Hubble Space Telescope's Cosmic Origins Spectrograph, a spectrometer that works in the ultraviolet bands; or x-ray telescopes, like Chandra and XMM-Newton, that can produce x-ray spectra. Successful detections (such as those in the Sculptor Wall) can feed back into the models, providing vital empirical evidence for the abundance and distribution of this elusive material. This is a perfect example of theory and observations working together to progress our knowledge.

There are tensions between observations and numerical models. I mentioned above that N-body simulations are limited by resolution: you can model a large chunk of the universe containing millions of galaxies, but not really model the galaxies themselves in a huge amount of detail. Alternatively you can chose to model a single galaxy at high resolution, but not simulate its large-scale environment simultaneously. Very large N-body simulations *have* been performed to examine the evolution of dark matter in individual galaxies, or rather, haloes, that are similar to the Milky Way. The technique is to take a large-volume simulation of the universe, like the Millennium Simulation, then identify a handful of galaxies that you want to simulate in more detail. Once you know the locations of these systems, you can run a new simulation, with the same physical model and initial conditions, but just focused on these individuals.

A recent project to do just this is called Aquarius, which has selected six examples of haloes formed in the Millennium that are thought to be analogous to our Milky Way. New N-body simulations have now been run that use up to 200 million dark-matter particles to represent each of these systems (one simulation has modelled one of the haloes at even higher resolution, with 1.5 billion particles). The results are beautiful – showing the intricate details in the dark-matter distribution in the halo – but troubling. One problem is that when you look at the structures of these galaxy haloes, similar to the Milky Way, you find a huge number of sub-structures within them. Sub-haloes are expected as part of the hierarchical nature of structure formation – just as a massive galaxy cluster halo contains sub-haloes (the population of individual galaxies within it), so a single galaxy halo contains further sub-haloes.

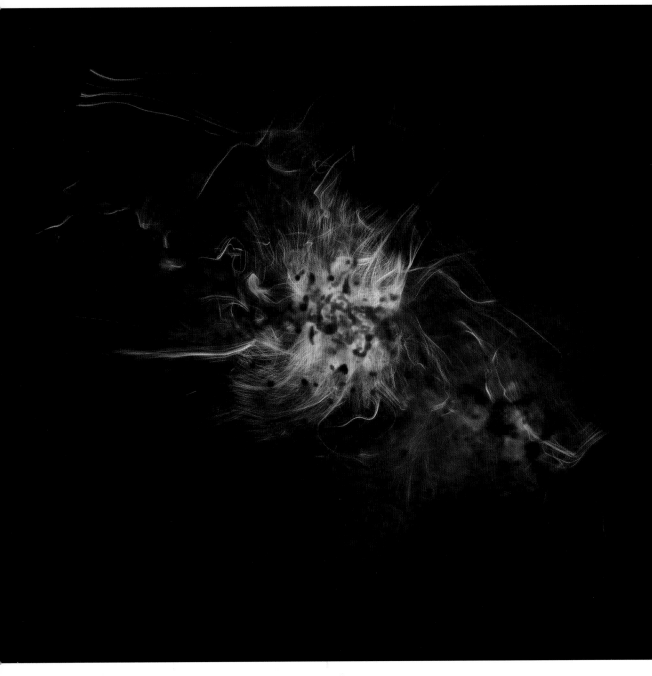

Some computer simulations also track the evolution of normal matter, as well as dark matter. This image shows a galaxy that has formed within a simulated universe, which has been primed with everything we think we know about cosmology and the physics of galaxy formation. Red colours and streams show cool gas flowing into a central nascent disc which is forming stars. Blue colours and streams show hotter gas emanating from the disc and forming a hot halo around the galaxy. Galaxy formation is really about the flow of gas into, within and out of galaxies: understanding that cycle is where much effort (both observationally and theoretically) is being spent right now.

We know this halo sub-structure exists, because galaxies like our own have obvious satellite galaxies – the LMC and SMC being the biggest ones for the Milky Way – around them. The problem is the sheer number of them that appear in the simulations. Our galaxy is not surrounded by thousands of dwarf satellites – at least, not that we can see – only a relative handful. This is called the 'satellite problem'.

One solution could lie in the baryon physics – the flow of gas and electromagnetic radiation in those haloes. Remember, the N-body simulation only shows us the evolution of dark matter, which we can't see directly in practice, just its gravitational effects (for now, at least). Perhaps these dark-matter satellites really are out there, surrounding and orbiting the galaxy but containing no stars or gas, like ghost towns in the suburbs. Is there a plausible physical explanation to back up this hypothesis? As we know, clumps of dark matter accrete baryons through gravity, pooling gas. However, it is possible to remove this gas by exerting a force on the baryons that can overpower the gravitational grip. As the strength of the gravitational grip is dependent on mass, it is easier to remove gas from low-mass haloes (like the satellites in the simulation) than it is from more massive ones (like the parent halo that the sub-structure is associated with).

Consider the formation of a galaxy like the Milky Way. A large halo forms from the accretion of smaller dark-matter clumps, gradually building up a massive halo surrounded by and embedded with myriad sub-haloes. At the same time, baryons – gas – collect within them. In the densest, central part of the halo we have the foundation of what will become the disc of the galaxy, and surrounding the disc like flies, proto-satellites begin to form.

At some point, star formation is initiated in the galaxy. For stars that form inside an individual sub-halo – a dwarf galaxy – it's possible that, after a few million years, when the first supernovae explode, they will blast out *all* of the gas in the sub-halo. In effect they have snuffed themselves out: the energy released by the supernovae is comparable to or greater than the gravitational binding energy of the dwarf. Star formation occurring in the developing disc also exerts pressure on the surrounding dwarfs by bathing them in stellar radiation and blasting them with the wind driven by supernovae and stars. If a central black hole starts to grow, more feedback energy can be released. Like an incoming tide washing away a sandcastle, gas can be removed from the sub-haloes. A few of the more massive satellites may keep some of their gas, forming visible companions that prevail to the present day, as evidenced by observation. This is just a hypothesis, of course – it could be that the simulations are wrong and that they are producing *too much* sub-structure in the universe. This is a source of concern, although not panic, in the Lambda-CDM paradigm, reminding us that we have work to do.

Perhaps the nature of dark matter is different from that described by our model; the luxury of simulations is that you can rebuild the universe with a new set of rules. If the dark matter is made slightly warmer, for example, then the same Aquarius simulation does not make as many sub-haloes. It is more closely matched to the number of satellites actually observed in nature, and this could be an important clue. Until we have empirical data on the properties of dark matter itself (what sort of particle is it, for example), or observations that might indicate the accuracy of one model or the other (the ability to somehow detect these putative barren dark satellites, perhaps through their gravitational interaction with stars in the disc of the galaxy, for example), this remains one of the mysteries of galaxy formation. It's what makes this game exciting: we have something interesting to solve!

An image showing the distribution and temperature of gas in a large chunk of a simulated universe: white, brighter hues are regions where the gas is hot (millions of degrees), whereas red/orange colours denote cooler gas. By studying the flow of gas into and out of galaxies in computer simulations we can hope to better understand how galaxies have formed and evolve through the interplay between gravitational forces and the feedback that occurs as stars and black holes churn out vast quantities of energy into their local environments.

Theory and simulations have allowed us to explore, using our best guess for the physics involved, how galaxies formed from the primordial chaos. What I find most remarkable about the process of galaxy formation is the way in which highly complex systems, with structure and order on a huge range of scales, have evolved from initial conditions through a set of relatively simple rules that govern the motion and behaviour of the most basic elements: the particles of matter. Returning to the beginning of our story, perhaps the most graphic illustration of this is in the beautiful spiral arms of galaxies in the local universe. How did those pinwheels of light form and persist?

The formation of galaxies

We have talked about how galaxies formed initially from the collapse of density fluctuations in a smooth sea of matter. Gas streaming into one of these 'over-densities' can form a flattened, rotating disc because the overall clump of matter from which the proto-galaxy collapsed had bulk angular momentum, it was twisting and turning due to gravitational tidal torques and interactions within the larger-scale matter distribution. As the halo

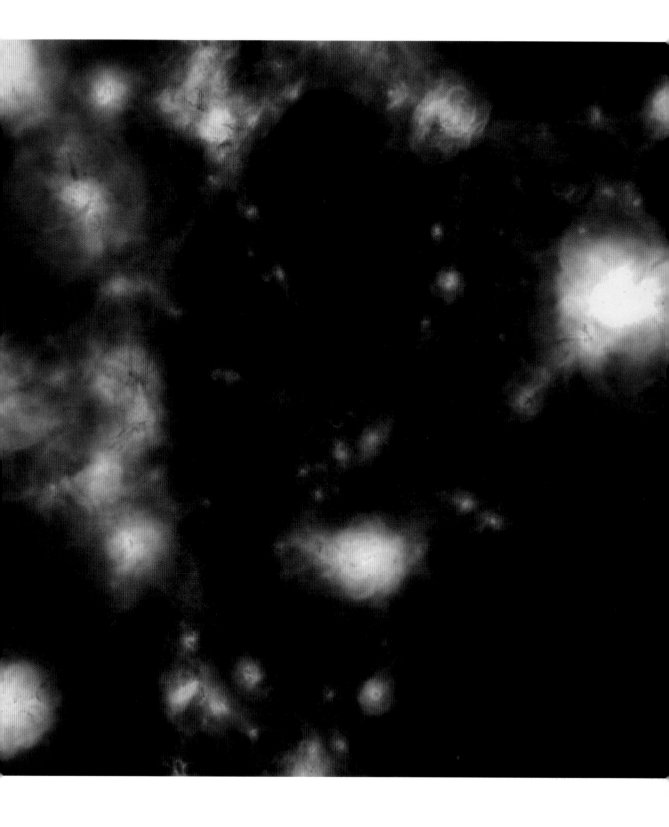

collapses through gravity, this angular momentum is conserved (one of the basic rules of physics, familiar from school), and the galaxy 'spins up' as it contracts and reduces in radius. With this spin, baryons settle into a rotationally supported disc, like spinning pizza dough.

Centrifugal forces act on the gas in the spinning disc, pushing radially outwards and holding it up against gravity, preventing it from collapsing down into a single clump. Provided the angular momentum is not lost, this allows a galaxy to retain a disc over long periods of time. This is a canonical picture – the exact physics of disc formation is a bit more complicated – but it is the basic picture that explains the existence of 'disc galaxies' like our own. Since there is differential rotation in the disc (it is not a rigid body; the gas is relatively loosely bound together within the disc itself), spiral arms can form from density perturbations that propagate through the rotating disc. Gas and new stars that are in orbit around the centre of the galaxy can accumulate in patches as the density wave passes. The origin of the wave could be a random gravitational disturbance, or perturbation of the disc caused by a nearby halo, or the accretion of a satellite, for example.

The analogy often given is the example of cars that form a localized, slowly propagating traffic jam on a motorway, perhaps caused by a slow-moving vehicle. As the faster-moving cars flow past (some overtaking, some stuck behind), we experience a brief bunching of traffic that can propagate along the road. The slow-moving vehicle is like the density wave, and as it propagates through the disc, the gas and stars can get bunched up around it. However, since the disc is also rotating differentially, the bunchings of gas and stars get wound into a spiral pattern. These arms stand out even more because the increased density of gas results in an increased rate of star formation (indeed, the density waves can *trigger* the star formation by inducing GMCs to collapse), so in spiral galaxies we see that the spiral arms are full of bright blue stars and patches of ionized gas emission.

If the angular momentum of a disc can be dissipated, for example through the merger of two galaxies, then the rotational support can be lost and the system will evolve into something more like a bulge or elliptical galaxy, where the shape is driven by the fact that the stars are moved onto random, rather than circular, orbits around a common mass. These are called 'pressure supported' or 'dispersion dominated' systems. Elliptical galaxies are such systems, the product of build-up via a large number of mergers early in their history that has erased any ordered rotation the constituent stars may once have had. Incidentally, this violent early history may also explain their old stellar ages – if the mergers happened early in ellipticals' evolution (which, remember, form in the densest environments), then this might have triggered a large amount of star formation, exhausting most of the gas very quickly.

The formation of the bulge of a disc galaxy, which is not rotationally supported, is a slightly contentious issue. Some bulges might form like miniature ellipticals through minor merging with a bunch of smaller systems early in the history of the galaxy. This 'classical' bulge could then acquire a disc by accreting fresh gas in the manner described above. Alternatively, bulges might grow over time as gas and stars are funnelled and transported to the hub of the galaxy through dynamical instabilities that cause the loss of angular momentum. Once driven to high densities, nuclear star formation can be triggered, depositing stellar mass in the core, puffing up in what is termed a 'pseudo-bulge'. It is likely that galaxy bulges form through both processes: there are many roads to Rome.

Forward steps

The field of galaxy evolution is one of the most dramatic illustrations of the power of the application of science as a tool for understanding the world. We began with a simple question: what were the faint, fuzzy nebulae between the stars? We had ideas, some wrong, some right, but it was the simple, careful and meticulous observation of the sky that held the key to the profound realization that these patches of light were external, independent star systems separated from our own by an almost inconceivable gulf of space. We then found out that there are different types of galaxy, and that they are not static, but in motion relative to us, moving away from the Earth with a speed that increases with distance. In a remarkable period at the start of the twentieth century, our very *concept* of the universe changed. The cosmos was far bigger, and richer, than any of our ancestors had dreamed. We moved forward as a species.

Now, at the beginning of the twenty-first century, and with the passage of several generations of astronomers since those early steps, we have dramatically built on this picture by looking deeper, farther and in ever more detail at the galaxies. We have made maps of the entire sky at many different wavelengths of light, detecting millions of galaxies and mapping their distribution into groups and filaments and clusters (and even clusters of clusters). We have imaged deep, small, patches of sky – like keyholes in time – to probe right back to when the universe was in its infancy, when it was approximately a tenth of its present size, half a billion years after the Big Bang. We have measured how galaxies have changed over cosmic history, determined their chemistry and composition and shape and dynamics, and framed this in a single theoretical model that seems to accurately describe the large-scale evolution of the universe as a whole.

And yet it feels as though we've only just hit our stride. There is so much more to know. The next two decades will bear witness to significant advances that build on and surpass everything we have learned so far, both in observation and theory.

The next generation of telescopes is now being designed and built, with a simple goal of seeing more clearly than we have before. We've already mentioned the Atacama Large Millimeter array or ALMA, now close to completion in northern Chile, which will, with its 50 12-metre-wide antennae spread across 16 kilometres of the Atacama Desert, allow us to measure the chemistry and dynamics of the cold interstellar media, the cold gas and dust, in star-forming galaxies across cosmic time. ALMA will fill in the 'missing link' of galaxy evolution – observations of the molecular fuel from which all the stars we see around us were formed. Although telescopes exist that allow us to detect the gas in distant galaxies, we're currently limited to only the most luminous galaxies; those with the most gas. ALMA is a great step forward, capable of detecting gas in a galaxy like the Milky Way seen just a few billion years after the Big Bang. It represents uncharted territory for galaxy evolution studies.

Still in the planning stage, the Square Kilometer Array (SKA) will be a radio telescope with a collecting area of a *million* square metres, sensitive enough to 'detect the radio signal from an airport radar on a planet fifty light years distant'. Once complete, SKA will represent a watershed moment in radio astronomy. SKA is not yet built, but two 'pathfinder' radio telescopes – prototypes of SKA technology – are already being built: MeerKAT and ASKAP, based in South Africa and Australia respectively. These pathfinders alone will be the most powerful radio telescopes ever built, capable of detecting pretty much every star-forming galaxy and active galactic nucleus over nearly half of the history of the universe. Untold riches await.

In the optical and near-infrared bands, plans are afoot to build 'Extremely Large' telescopes that dwarf the largest we currently have, increasing their collecting area into giants with mirrors 30 to 50 metres across. These

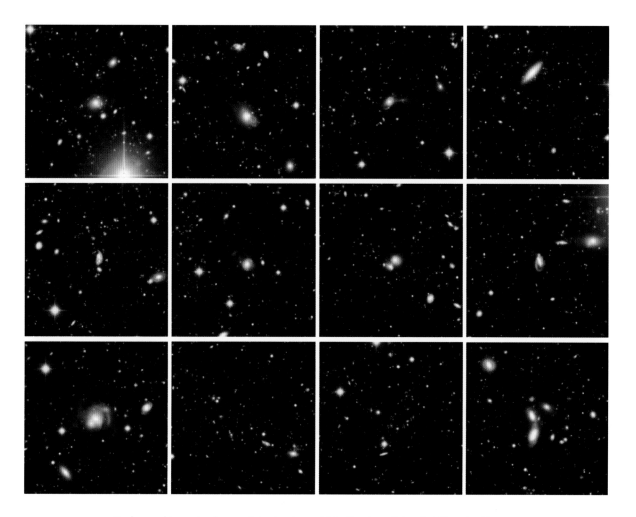

Each panel is centred on a distant galaxy within the deep Extended Chandra Deep Field South survey region. Several of these galaxies show evidence of gravitational interactions and mergers, with disturbed morphologies and stellar streams: a common process shaping the evolution of galaxies. Nearly every speck of light in these images represents emission from countless, even more distant galaxies. Extragalactic astronomers use deep survey fields such as this to study large samples of distant galaxies, utilizing the fact that light itself takes an appreciable amount of time to traverse the universe to our telescopes and detectors here on Earth to literally see back into the past. By patiently observing the universe in this way we have built up a comprehensive, but incomplete, picture of how galaxies have formed and evolved over nearly 14 billion years of cosmic history. We have learned much, but great discoveries still await.

gargantuan light buckets will allow us to detect and measure the stars in even more distant galaxies, with sensitivities far beyond anything that has come before. There will also be new 'synoptic' surveys, such as the planned Large Synoptic Survey Telescope (LSST), which will repeatedly image the majority of the sky, not only building-up a large, deep image that will detect millions of galaxies, but creating a kind of movie of the universe, each imaging pass adding a frame. This will allow LSST to hunt for supernovae and other transient phenomena that blink in and out of the picture as it builds its long-exposure image of the universe over the span of a decade.

In space, we will – hopefully – see the successor to the Hubble Space Telescope, the James Webb Space Telescope (named after James Webb, the second NASA administrator, influential in the landmark Apollo mission). Placed in space a million miles from Earth, the JWST will operate in the near- and mid-infrared parts of the spectrum, collecting light with a 6.5 metre mirror made up from a mosaic of segments that will deploy once the telescope is in space. JWST, when it opens its eyes, will provide a glimpse into the Dark Ages, detecting galaxies close to the time when stars first shone. Other satellites are also planned: Gaia will map the locations of half a billion stars in our galaxy, and Euclid will map the entire sky at near-infrared wavelengths and detect millions of distant star-forming galaxies, the statistical distribution of which will provide information on the nature of dark energy. Astronomers are continuously dreaming up new experiments and missions, some of which may only be possible many decades from now. The fates

of these dreams are dependent on the availability of requisite technology, the volatile swings in economic climate and political and public feeling about science investment and international collaboration.

Aside from the big projects taking us forward, there will of course be the continued development of new instrumentation for our existing telescopes – new cameras and spectrographs offering improved sensitivity and new observational techniques for example. At the same time as these advances in observational technology, computational power continues to increase and the costs of hardware decrease, allowing more sophisticated, higher-resolution and larger simulations with which to explore and probe our models, and with which to compare and help interpret the flow of empirical data that is increasing from a river to a torrent.

With these efforts, our World Model will continually be refined. If, at the start of the twentieth century, our model of the universe was a misshapen lump

This collection of pixels represents light from one of the most distant, and therefore earliest, galaxies known. As we look further, astronomy gets harder and harder as the signal from distant galaxies is dimmed by the – literally – astronomical distances between Earth and these far-flung cosmic sources. By detecting them, we are seeing into the past, as the light we are now detecting left the galaxies billions of years ago.

of marble, by the middle of the twenty-first century it will be a David. It has never been more exciting to be an extragalactic astronomer than it is now. With the promise of new discoveries ahead, we continue the adventure that, through the constant attrition of science, is revealing the mysteries of the universe. The statue beneath the stone.

Distance Scale

1 milliparsec (mpc) = 0.001 pc
1 kiloparsec (kpc) = 1000 pc
1 Megaparsec (Mpc) = 1'000'000 pc
1 Gigaparsec (Gpc) = 1'000'000'000 pc

Values in parentheses correspond to a scale where the Earth–Sun distance is 1 millimetre

Distance from Earth to Moon: 0.00001 mpc (0.003 mm)

Distance from Earth to Sun: 0.005 mpc (1 mm)

Diameter of solar system (to heliopause): 1 mpc (20 cm)

Distance from Sun to Proxima Centauri: 1.3 pc (270 m)

Distance from Sun to Orion nebula: 410 pc (85 km)

Diameter of Betelgeuse: 0.05 mpc (1 cm)

Diameter of Orion nebula: 6 pc (1.2 km)

Distance from Sun to 47 Tucanae: 5.1 kpc (1,050 km)

Diameter of 47 Tucanae: 37 pc (7.6 km)

Distance from Sun to centre of Milky Way: 8 kpc (1,650 km)

Thickness of Milky Way stellar disc: 300 pc (60 km)

Diameter of Milky Way disc: 30 kpc (6,200 km)

Radius of Milky Way bulge: 5 kpc (1,000 km)

Distance from Sun to Magellanic Clouds: 50 kpc (10,300 km)

Distance from Milky Way to Andromeda galaxy: 780 kpc (160,000 km)

Approximate diameter of Local Group: 3 Mpc (620,000 km)

Distance from Milky Way to Virgo cluster: 16.5 Mpc (3.4 million km)

Diameter of Virgo cluster: 2 Mpc (400'000 km)

Distance from Milky Way to Coma: 100 Mpc (21 million km)

Diameter of Coma cluster: 6 Mpc (1.2 million km)

Co-moving distance* to the most distant galaxy identified: 10 Gpc (2 billion km)

Light travel time from most distant galaxy identified: 13.3 billion years

* *Note: the co-moving distance is the distance to an object taking into account the expansion of the universe over time.*

Glossary

ABSORPTION LINE: a dip in the observed spectrum of a star or galaxy due to the absorption of light at a very specific frequency by a certain element. Like emission lines, the exact wavelength or frequency of the absorption line is determined by the energy of the absorbed photon. The spectrum of the Sun reveals many dark absorption features that correspond with the presence of metals such as calcium and sodium, as well as absorption by the hydrogen atom. Emission and absorption lines can reveal information about the chemical composition and dynamical state of a galaxy, as well as its redshift.

ACTIVE GALACTIC NUCLEUS: the name given to the core of a galaxy that contains an actively accreting supermassive black hole. This process releases copious amounts of energy as material is heated during its passage near (and into) the black hole.

ADAPTIVE OPTICS: a method to compensate for the distorting effect of the Earth's atmosphere, which limits the resolution of images that can be taken from the ground.

ATOMIC GAS: gas comprising single atoms, such as hydrogen.

BARYONS: the name given to the class of matter that characterizes 'normal' material, including atoms. By mass, baryons only account for 5 per cent of the universe.

CEPHEID VARIABLE: a star that pulses in brightness over a period of days to weeks. The period of the variation is directly related to the average luminosity of the star, so Cepheid variable stars can be used as 'standard candles' for distance measurements.

CHARGE-COUPLED DEVICE (CCD): a workhorse instrument of astronomy, taking the place of a photographic plate for recording light. CCDs are constructed from semi-conducting material, and are essentially a two-dimensional array of individual detectors (pixels) that generate a small electrical charge when exposed to light. The amount of charge measured can be converted to incoming flux, thus providing images of astronomical sources. CCDs operate in the ultraviolet-optical-near-infrared bands of the electromagnetic spectrum.

CLUSTER OF GALAXIES: the name given to the huge congregations of up to thousands of individual galaxies residing in the most massive dark-matter haloes. The total mass of these systems can be up to a million billion times the mass of the Sun. The intracluster medium of clusters (ICM) is filled with a hot plasma of gas many millions of degrees in temperature, and this can affect the galaxies that are moving rapidly within it, for example through ram-pressure stripping.

DARK ENERGY: the name given to the substance or physical mechanism that gives rise to the observed acceleration of the rate of expansion of the universe, as revealed by, for example, trends in the observed brightness of distant supernovae with redshift. There are several theoretical ideas for what dark energy might be, but so far none have been confirmed. Since there is an equivalence between energy and mass, dark matter contributes to the total mass of the universe, and represents about two-thirds of the total. Both dark matter and dark energy are beyond the standard model of physics.

DARK MATTER: the name given to the substance that appears to constitute about a quarter of the mass of the universe, but does not interact with normal baryonic matter in any other way apart from the gravitational force. We observe no electromagnetic emission from dark matter and have not directly detected it. However, we can see its effects in, for example, the rotation curves of galaxies (which are governed by the distribution of mass in a galaxy) and gravitational lensing. Empirically understanding the nature of dark matter is a key goal of modern astronomy. In the current model, the particles that make up dark matter are thought to move slowly compared with the speed of light, so it is referred to as 'Cold Dark Matter' (CDM).

DWARF GALAXY: a low-mass galaxy, often companion to a larger galaxy. Dwarf galaxies are usually irregular in shape and can be rapidly forming stars.

ELECTROMAGNETIC RADIATION: visible light is part of the electromagnetic spectrum, but this extends to lower and higher wavelengths, corresponding to different energies of radiation. Different physical processes can release or give rise to emissions of electromagnetic radiation right across the spectrum, from radio waves to gamma rays.

EMISSION LINE: a spike in the observed spectrum of a gas cloud, star or entire galaxy corresponding to a specific transition of electrons between different energy states in an atom. The energy of the transition determines the wavelength or frequency of the line. The gas clouds around newly forming stars glow brightly with the light of ionized hydrogen, and are called HII regions: the gas is energized by the light blazing from the young, massive stars that have ignited.

EPOCH OF RECOMBINATION: the time in the universe's history when neutral hydrogen (a proton and electron bound together) formed from the hot plasma of protons and electrons. Occurred around 350,000 years after the Big Bang.

EPOCH OF REIONIZATION: the time in the universe's history when the first stars ignited, which then started to ionize the mainly neutral intergalactic gas. Thought to have occurred around 500,000 years after the Big Bang, but over an extended period of time.

FAR-INFRARED/SUBMILLIMETRE LIGHT: part of the electromagnetic spectrum beyond the mid-infrared, with wavelengths of 100 microns to a millimetre. Astrophysical sources include cool dust (few tens of degrees).

FEEDBACK: processes that release energy into the interstellar (and intergalactic) medium that can modify or otherwise affect the gravitational collapse of gas, thus controlling star formation. Feedback processes can include stellar winds blown from the surfaces of hot stars, the explosive release of energy in supernovae or powerful jets emerging from an actively accreting supermassive black hole at the heart of a massive galaxy. Feedback is a critical component in our current model of galaxy evolution, limiting the runaway gravitational collapse of gas into stars.

FLUX: the observed flow of energy through a detector (for example, a CCD camera) from a distant source, measured in units of watts per square metre.

GRAVITATIONAL LENSING: the effect of distortion and magnification of background sources along the line of sight to a massive structure, like a cluster of galaxies, due to the gravitational warping of space-time. Lensing can not only allow us to study very distant galaxies in great detail, but to measure the total mass (including dark matter) of the object acting as the lens.

HUBBLE TYPE: a classification scheme that identifies galaxies by their morphological type. The main Hubble types are elliptical, lenticular, spiral and barred spiral.

INTERSTELLAR: the space between stars within a galaxy.

INTERGALACTIC: the space between galaxies.

INTRACLUSTER: the environment between galaxies within a cluster of galaxies.

IONIZATION: the process by which an electron in an atom can be removed when a photon of sufficient energy is absorbed by the atom.

LOCAL GROUP: name given to the local volume of space around the Milky Way containing several tens of galaxies, including the galaxy M31 (Andromeda).

LUMINOSITY: the total amount of energy released from an object (such as a galaxy), measured in units of watts.

MAIN SEQUENCE: the name given to the tight locus in a plot of luminosity versus colour (the Hertzsprung-Russell diagram) for stars of different masses, representing the phase of hydrogen burning in stellar evolution.

MID-INFRARED LIGHT: part of the electromagnetic spectrum beyond the near-infrared, with wavelengths of a few to tens of microns. Astrophysical sources include hot dust (hundreds of degrees).

MOLECULAR GAS: gas clouds comprising mainly hydrogen molecules: two hydrogen atoms bonded together. Giant clouds of molecular gas (GMCs) are the places where new stars can form.

N-BODY SIMULATION: a computer simulation that models the gravitational evolution of structure using an ensemble of particles in three-dimensional space, with each particle representing a certain mass. The gravitational forces between all the particles are calculated, their accelerations applied and the simulation progressed to the next time step. Large N-body simulations (sufficient to model a representative volume of the universe at adequate resolution) are computationally expensive, requiring a supercomputer to run. The Millennium Simulation is an example of a large N-body simulation that has studied the evolution of dark matter in the universe assuming our current cosmological model (Lambda-CDM).

NEAR-INFRARED LIGHT: part of the electromagnetic spectrum with wavelengths (frequencies) longer (lower) than optical/visible light. Astrophysical sources include older, low-mass and cooler stars.

NEUTRAL GAS: a phase of gas where the atoms have not been ionized.

NUCLEOSYNTHESIS: the name given to the process of formation of the nuclei of elements. Light elements (hydrogen, helium, lithium) were formed shortly after the Big Bang, but heavier elements have been formed in stars, either during stellar burning or – for some stars – in their explosive deaths (supernovae).

OPTICAL OR VISIBLE LIGHT: the part of the electromagnetic spectrum that we can see with our eyes. Astrophysical sources include typical stars (like our Sun) and light from ionized hydrogen gas.

PARALLAX: the apparent change in the position of an object relative to a fixed background when viewed along different sightlines.

PARSEC: short for 'parallax second', a basic unit of distance in astronomy, equivalent to 3.26 light years, or 30,000 billion kilometres. One megaparsec is a million parsecs.

PHOTON: a packet of electromagnetic radiation. Light can be thought of as a stream of discrete photons, each of which is described by a particular frequency or wavelength. The energy of a photon is directly proportional to its frequency and inversely proportional to its wavelength.

QUASAR: quasi-stellar object – a class of galaxy where the energy emission is dominated by the activity around the central black hole. Quasars appear as points of light on the sky (like stars), with little of the 'host' galaxy visible. Because they are so bright, quasars can be seen over vast cosmic distances. Quasars represent an important stage in the formation of massive galaxies.

RADIO GALAXY: a galaxy (often a massive elliptical) that emits copious amounts of energy in the radio part of the electromagnetic spectrum. The radio waves originate from electrons that are accelerated in strong magnetic fields. Some galaxies exhibit spectacular bipolar jets that punch out of the galaxy into intergalactic space, originating from a central active nucleus: a supermassive black hole at the core of the galaxy that is accreting material.

RED GIANT: the late stage of stellar evolution after hydrogen has been burned, and the outer atmosphere of the star becomes highly extended.

REDSHIFT: the name given to the observed increase in wavelength (or equivalent decrease in frequency) in light emitted by a body that is moving away from our frame of reference. The cosmological redshift arises because the universe is expanding: light emitted from distant galaxies was released when the universe was a fraction of its current size and the universe has expanded in the meantime, so the galaxies appear to be receding.

SPECTRAL TYPE: classification scheme for stars on a scale of hot/luminous/blue to cool/faint/red. The basic sequence runs (from hot to cool) O, B, A, F, G, K, M; the Sun is a G-type star.

SPECTRUM/SPECTROSCOPY: light from a source (be it the Sun or a galaxy) can be dispersed into its constituent frequencies. The effect can be seen in rainbows, where the mixture of violet-to-red light that makes up sunlight is split as it passes through refractive raindrops. Spectroscopy can be used to learn about the amount of energy emitted at different frequencies, and therefore provide clues about the composition and physics of a particular system.

SURFACE OF LAST SCATTERING: at recombination, photons bouncing around within the hot plasma were released, free-streaming across the universe. We detect these as the Cosmic Microwave Background, and this is the most distant light we can see. Satellites like WMAP and Planck have mapped the CMB and revealed the variations in its temperature corresponding to density fluctuations that represent the seed-points of galaxy formation.

ULTRAVIOLET LIGHT: part of the electromagnetic spectrum with wavelengths (frequencies) shorter (higher) than optical/visible light. Astrophysical sources include hot, young stars.

WHITE DWARF: compact remnant formed at the end of a star's life (often in the centre of an extended nebula representing the expelled outer layers of the star).

Bibliography

Banks, Iain M., *The State of the Art* (London, 1993)

Coles, Peter, *Cosmology: A Very Short Introduction* (Oxford, 2001)

Greene, Brian, *The Fabric of the Cosmos* (Harmondsworth, 2005)

Gribbin, John, *Galaxies: A Very Short Introduction* (Oxford, 2008)

—, *Stardust* (Harmondsworth, 2009)

Hawking, Stephen, *A Brief History of Time* (London, 1998)

Hubble, Edwin, *Realm of the Nebulae* (Silliman Memorial Lecture Series) (New Haven, CT, 2013)

Jones, Mark H., and Robert J. Lambourne, eds, *An Introduction to Galaxies and Cosmology* (Maidenhead, 2004)

Longair, Malcolm, *The Cosmic Century* (Cambridge, 2006)

Mo, Houjun, Frank van den Bosch and Simon White, *Galaxy Formation and Evolution* (Cambridge, 2010)

Moore, Sir Patrick, *Philip's Guide to the Night Sky* (London, 2013)

Rees, Martin, *Our Cosmic Habitat* (Princeton, NJ, 2003)

—, ed., *Universe* (Oxford, 2012)

Rowan-Robinson, Michael, *Night Vision* (Cambridge, 2013)

Sagan, Carl, *Cosmos* (New York, 2013)

Sanders, Robert H., *Revealing the Heart of the Galaxy: The Milky Way and Its Black Hole* (Cambridge, 2013)

Scharf, Caleb, *Gravity's Engines* (New York, 2012)

Smoot, George, and Keay Davidson, *Wrinkles in Time* (New York, 1993)

Sparke, L. S., and J. S. Gallagher III, *Galaxies in the Universe: An Introduction* (Cambridge, 2007)

Sparrow, Giles, *Constellations: A Field Guide to the Night Sky* (London, 2013)

—, *Cosmos: A Journey to the Beginning of Time and Space* (London, 2007)

—, *Hubble: Window on the Universe* (London, 2010)

Tyson, Neil Degrasse, and Donald Goldsmith, *Origins: Fourteen Billion Years of Cosmic Evolution* (New York, 2006)

Weinberg, Steven, *The First Three Minutes: A Modern View of the Origin of the Universe* (New York, 1993)

Acknowledgements

Thanks to Fabian Walter for permission to print the THINGS image of NGC 268, to Alice Danielson and Mark Swinbank for providing the VLT/FORS 2d spectrum of a distant galaxy from the ZLESS survey and to Volker Springel for permission to use visualizations of the Millennium Simulation. I would also like to thank Tammy Hickox and Tim Geach for excellent advice. Finally, and most importantly, I could not have written this book without the love, support and patience of my wife Kristen and our daughter Sophie. You are the centre of my galaxy.

Photo Acknowledgements

ALMA (ESO / NAOJ / NRAO), NASA / ESA: pp. 186, 194 (J. Hodge et al., A. Weiss et al., NASA Spitzer Science Center); Michael Blanton and the Sloan Digital Sky Survey (SDSS) collaboration, http://www.sdss.org: p. 97; ESA /Herschel / PACS & SPIRE Consortium, O. Krause, HSC, H. Linz: p. 180; ESA / Hubble & NASA: p. 159; ESA, LFI and HFI consortia: p. 30; ESO: pp. 6 (José Francisco Salgado), 10–11, 26, 49, 54–55 (S. Guisard), 68–69, 75 (S. Brunier), 21 (S. Brunier / S. Guisard [www.eso.org/~sguisard]), 19, 24, 29 (R. Schoedel), 44 (P. Grosbøl), 57, 62 (Y. Beletsky), 85, 103 (S. Gillessen et al.), 110–111 (J. Pérez), 116, 132–133, 138 (J. Emerson / VISTA / Cambridge Astronomical Survey Unit), 119, 124, 125, 126 (F. Comeron), 142 (C. Malin), 157, 160, 247; ESO / APEX (MPIfR / ESO / OSO) / A. Hacar et al. / Digitized Sky Survey 2: p. 127 (Davide De Martin); ESO / M.-R. Cioni / VISTA Magellanic Cloud Survey: p. 14 (Cambridge Astronomical Survey Unit); ESO / Digitized Sky Survey 2: p. 161; ESO / M. Gieles: p. 139 (Misha Shirmer); ESO / INAF-VST / OmegaCAM: pp. 13 (A. Grado / INAF-Capodimonte Observatory), p. 155 (OmegaCen / Astro-WISE / Kapteyn Institute); ESO / SOAR / NASA: p. 170; ESO / VVV Survey / D. Minniti: p. 28 (Ignacio Toledo, Martin Kornmesser); ESO / WFI (Optical); MPIfR / ESO / APEX / A. Weiss et al. (submillimetre); NASA / CXC / CfA / R. Kraft et al. (x-ray): p. 88; J. Geach: p. 93; J. Geach / R. Crain: pp. 240, 243; NASA: pp. 60, 104; NASA, N. Benitez (JHU), T. Broadhurst (The Hebrew University), H. Ford (JHU), M. Clampin (STScI), G. Hartig (STScI), G. Illingworth (UCO / Lick Observatory), the ACS Science Team and ESA: p. 219; NASA, ESA, S. Beckwith (STScI) and the HUDF Team: p. 17; NASA, ESA, CXC, SAO, the Hubble Heritage Team (STScI / AURA), and J. Hughes (Rutgers University): p. 112; NASA / CXC / IoA / A. Fabian et al.: p. 167; NASA, ESA, S. Baum and C. O'Dea (RIT), R. Perley and W. Cotton (NRAO / AUI / NSF), and the Hubble Heritage Team (STScI / AURA): pp. 164–165; NASA, ESA, S. Beckwith (STScI), and the Hubble Heritage Team (STScI / AURA): pp. 42–43; NASA, ESA and K. Cook (Lawrence Livermore National Laboratory, USA): p. 36; NASA, ESA, CXC, SAO, the Hubble Heritage Team (STScI / AURA), and J. Hughes (Rutgers University): p. 112; NASA, ESA, and G. Canalizo (University of California, Riverside): p. 101; NASA / ESA and ESO: p. 189; NASA, ESA, M. Postman and D. Coe (STScI) and the CLASH Team: p. 249; NASA, H. Ford (JHU), G. Illingworth (UCSC / LO), M. Clampin (STScI), G. Hartig (STScI), the ACS Science Team, and ESA: pp. 195,

198; NASA, ESA, and P. Goudfrooij (STSCI): pp. 46–47 top; NASA, ESA, and the Hubble Heritage Team (STSCI / AURA): pp. 26–27 (R. Corradi [Isaac Newton Group of Telescopes, Spain]) and Z. Tsvetanov (NASA), 37 (S. Smartt [Institute of Astronomy]) and D. Richstone [U. Michigan]), 38–39 (P. Knezek [WIYN]), 46–47 (bottom), 48 (W. Keel [University of Alabama, Tuscaloosa]), 90–91 (D. Carter [Liverpool John Moores University] and the Coma HST ACS Treasury Team), 114 (R. Corradi [Isaac Newton Group of Telescopes, Spain] and Z. Tsvetanov [NASA]), 156 (K. Cook [Lawrence Livermore National Laboratory]), 157, P. Goudfrooij (STSCI), 173 (J. Blakeslee [Washington State University]); NASA, ESA, and the Hubble Heritage Team (STSCI / AURA)-ESA / Hubble Collaboration: pp. 34 (R. Chandar [University of Toledo]) and J. Miller (University of Michigan]), 41 (M. Crockett and S. Kaviraj [University of Oxford], R. O'Connell [University of Virginia], B. Whitmore [STSCI], and the WFC3 Scientific Oversight Committee), 50, 158, 159 (R. O'Connell [University of Virginia] and the WFC3 Scientific Oversight Committee), 51 (M. West [ESO, Chile]), 192, 193, 193 (B. Whitmore [Space Telescope Science Institute]),196–197, 199, 206–207 (J. Gallagher [University of Wisconsin]), M. Mountain (STSCI), and P. Puxley (National Science Foundation); NASA, ESA, the Hubble Heritage Team (STSCI / AURA)-ESA / Hubble Collaboration, and A. Evans (University of Virginia, Charlottesville / NRAO / Stony Brook University): pp. 202, 203; NASA, ESA, and the Hubble SM4 ERO Team: p. 204; NASA, ESA, the Hubble SM4 ERO Team, and ST-ECF: p. 220; NASA, ESA, and J.-P. Kneib (Laboratorie d'Astrophysique de Marseille): pp. 174–175; NASA, ESA, G. Kriss (STSCI), and J. de Plaa (SRON Netherlands Institute for Space Research): p. 102 (B. Peterson [Ohio State University]); NASA, ESA, D. Lennon and E. Sabbi (ESA / STSCI), J. Anderson, S. E. de Mink, R. van der Marel, T. Sohn, and N. Walborn (STSCI), N. Bastian (Excellence Cluster, Munich), L. Bedin (INAF, Padua), E. Bressert (ESA), P. Crowther (University of Sheffield), A. de Koter (University of Amsterdam), C. Evans (UKATC / STFC, Edinburgh), A. Herrero (IAC, Tenerife), N. Langer (AifA, Bonn), I. Platais (JHU), and H. Sana (University of Amsterdam): pp. 128–129; NASA, ESA, and M. Livio and the Hubble 20th Anniversary Team (STSCI): p. 137; NASA, ESA, R. O'Connell (University of Virginia), F. Parescue (National Institute for Astrophysics, Bologna, Italy), E. Young (Universities Space Research Association / Ames Research Center) and the WFC2 Science Oversight Committee, and the Hubble Heritage Team (STSCI / AURA): p. 22; NASA, ESA, M. Regan and B. Whitmore (STSCI), R. Chandar (University of Toledo), S. Beckwith (STSCI), and the Hubble Heritage Team (STSCI / AURA) p. 35; NASA, ESA, A. Riess (STSCI / JHU), L. Macri (Texas A&M University), and the Hubble Heritage Team (STSCI / AURA): p. 40; NASA, ESA, and E. Sabbi (ESA / STSCI): p. 130; NASA, ESA, and R. Sharples (University of Durham): p. 205; NASA / ESA, N. Smith (University of California, Berkeley), and the Hubble Heritage Team (STSCI / AURA): pp. 134–135; NASA, ESA, ESO, CXC & D. Coe (STSCI) / J. Merten (Heidelberg / Bologna): p. 171; NASA / JPL-Caltech: pp. 78–79, 181, 182, 183, 184; NASA / JPL-Caltech / VLA / MPIA: p. 145; NASA / JPL-Caltech / WISE Team: p. 176; NASA / JPL-Caltech / the SINGS Team (SSC / Caltech): p. 185; NROA / AUI / NSF / GBT / VLA / Dyer, Maddalena & Cornwell x-ray: Chandra x-Ray Observatory; NASA / CXC / Rutgers / G. Cassam-Chenaï, J. Hughes et al., Visible light: 0.9-Curtis Schmidt optical telescope, NOAO / AURA / NSF / CTIO / Middlebury College / F. Winkler and Digitized Sky Survey: p. 113; N. A. Sharp,

NOAO / NSO / Kitt Peak FTS / AURA / NSF: pp. 82–83; Fabian Walter and the THINGS team (Walter et al. 2008): p. 144; Courtesy of Volker Springel and the Virgo Consortium: pp. 224, 236; ZLESS Consortium: p. 86.

Index

Page numbers in **bold** are Glossary entries